International Trade And Technology:

Models of Dynamic Comparative Advantage

Michael P. Claudon
Middlebury College

University Press of America™

University Press of America™
division of
R.F. Publishing, Inc.
4710 Auth Place, S.E., Washington, D.C. 20023

Printed in the United States of America

0–8191–0306–3

Preface

The purpose of this study is to obtain a better understanding of the product cycle model of international trade as developed by Raymond Vernon. The central problem to which the product cycle theory addresses itself is that of efficiently incorporating consideration of the influence of technological change on trade patterns into received comparative cost theory. Recent empirical evidence suggests that if orthodox trade theory is to reflect international trade patterns which are evolving, inclusion of the technological factor is essential.

Our central goal is to offer theoretical and empirical evidence supporting the position that the approach ennunciated by Vernon will prove to successfully incorporate the technological factor into received comparative cost models.

The nexus of the book involves the construction and testing of general equilibrium models of dynamic comparative advantage in both homogeneous and non-homogeneous new commodities whose production functions are distinguished by endogeneous technological change. This exercise expands upon the existing product cycle literature by indicating that consideration of non-homogeneous commodities -- which is dictated by real-world experience -- in no way distrubs the central conclusions of the product cycle theory.

A subsidiary contribution of the book is the incorporation of Kelvin Lancaster's "new approach" to consumer theory,[1] and the concept of consumer learning into dynamic microeconomic and macroeconomic models of consumer behavior given the introduction of one new commodity. With these models we are able to describe the equilibrium time path of quantities demanded for both homogeneous and non-homogeneous new commodities.

The book concludes with a presentation of the findings of our statistical analysis. While the evidence presented is of a very tentative nature, it does support acceptance of the

---

[1]Kelvin Lancaster, "Change and Innovation in the Technology of Consumption," American Economic Review, Papers and Proceedings, May, 1966; _______, "A New Approach to Consumer Theory, Journal of Political Economy, April, 1960.

product cycle approach to the inclusion into received comparative cost theory of consideration of the technological factor. Our tests indicate that the product cycle theory, as developed by Vernon and extended herein, is capable of distinguishing between mature commodities and new commodities. Based upon both cross-sectional and time-series tests on seven commodities, and fourteen countries, over the period 1950-1967, this model was not able to explain the external trade in mature commodities, while that of new commodities conformed rather closely to its expectations.

I wish to record my sincere gratitude to Dr. Bela Balassa of the Johns Hopkins University. He was a rich source of both valuable ideas and astute constructive criticism. I would like in addition to thank Dr. Trent Bertrand, also of the Johns Hopkins University, whose careful reading and criticism of earlier drafts helped me to bring my analysis into better focus.

Thanks are also due to Dr. Tom Sparrow. He was a valuable source of ideas and counsel for the subject matter of Chapter Two.

All errors, of course, remain the responsibility of the author.

CONTENTS

LIST OF TABLES

CHAPTER ONE

CHAPTER FOUR

CHAPTER FIVE

APPENDICES

LIST OF FIGURES

APPENDICES

Introduction

1. Theoretical Background

According to orthodox international trade theory as personified by the Hechscher-Ohlin-Samuelson (H.O.S.) model, both the composition and direction of international trade are determined solely by (static) differences in initial factor endowments, all other potential differences between economic agents (countries) having been assumed away. But, it has become increasingly difficult to reconcile the trade patterns predicted by this theoretical model with those which in fact have emerged, and which continue to evolve.

In attempting to reconcile theory and reality, economists have questioned both the H.O.S. model's static framework of analysis and the validity of the assumptions on which it is founded. Johnson observes that as a response to the apparent inconsistency between the theory and Leontief's original empirical findings,

> a number of economists concerned with the explanation of observed trade patterns, and especially with the explanation of the fact that a preponderant share of world trade is transacted between advanced industrial nations, have been emphasizing the importance in the explanation of trade patterns of factors abstracted from in the Hechscher-Ohlin model.[1]

The impact of technological change on trade patterns is one such factor. Economists have for some time been aware of the importance of technological change as a factor influencing international trade flows. In the 1950's, a theory of international trade emphasizing dynamic (technological) developments within an industry was hinted at by Kindleberger, Kravis, and Hoffmeyer.[2]

Kindleberger showed clearly that the United States' share of world exports of new and expanding commodities in the period 1913-1929 was far greater than its share in stagnating groups, whereas the U.K. and to a lesser extent, Germany, had a greater share in the less expanding or stagnating groups.[3]

In his stimulating analysis of U.S. trade patterns and their impact on the dollar shortage, Hoffmeyer found that from 1890-1854, the United States tended to specialize in what he termed research

Intensive commodities.[4]

His thesis is:

> Since the end of the nineteenth century the comparative advantage of the U.S. has increasingly been in the production of new and research intensive goods, whereas the exports of the rest of the world to the U.S. have been mainly traditional commodities with no great research intensity and which the U.S. to a large extent has been producing herself.[5]

Consideration of technological change and technological superiority as a source of comparative advantage has generated a novel, dynamic theory of international intercourse: the product cycle model of international trade. An early, rather crude statement of this theory was made by Hoffmeyer:

> We would expect that the U.S. surged forward especially in new products, but gradually lost its advantage in products that had become more traditional and less research intensive, assuming that new products are by and large more research intensive than other commodities. . . (I)t is typical for the U.S. to expand particularly in new and research intensive goods, whereas when such goods gradually become more traditional, the ordinary forces of the price system make themselves felt - in many cases with the effect that commodities are changed from export to import goods.[6]

A more developed statement of this theory has been enunciated by Posner.[7] He suggests that trade may be caused by technical changes and developments affecting particular industries, or creating new industries "because particular technical changes originate in one country, comparative cost differences may induce trade in particular goods during the lapse of time taken for the rest of the world to imitate one country's innovation."[8]

Vernon's 1966 Quarterly Journal of Economics paper is the first fully articulated statement of the product cycle theory.[9]

This approach modified the received body of international trade theory for the impact of the technological variable. The basic premise upon which Vernon's model is founded is that as the product cycle (life cycle) of any new commodity unfolds, its production passes through successive stages of standardization, and associated with these will be dynamic shifts in comparative advantage of predictable direction.

Vernon's model can be depicted graphically as in Figure 1.1. Following Vernon, three stages of standardization are identified: new, maturing (growth), and mature commodities.

New products have very high income elasticities of demand. Therefore, they can expect to be sold first in markets exhibiting the highest average incomes. Since the "U.S. market consists of consumers with an average income which is higher (except for a few anamolies like Kuwait) than that of any other national market. . . the chance to offer a new product responsive to wants at high income levels. . . would presumably first be apparent to someone in a position to observe the United States market."[10] But why is it the case that such new products also tend to be first-produced in the market for which they are designed?

The answer Vernon suggests is twofold. In early stages of production, the firm is likely to have a monopoly and therefore production costs are not likely to be a major locational factor. Secondly, during initial production periods (Stage I), the strongest locational forces are likely to be those of communication and external economies: "producers at this stage are particularly concerned with the degree of freedom they have in changing their inputs;. . . (and) the need for swift and effective communication on the part of the producer with customers, suppliers, and even competitors is especially high at this stage."[11]

But as the product passes out of Stage I and into Stage II (the Maturing Product), a certain degree of standardization obtains, and at this stage in a commodity's development, there is likely to be a shift in the location of production facilities intranationally, but more importantly internationally.

This expectation derives from declining requirements for flexibility and the increasing concern about production cost which accompanies the greater standardization in the manufacturing process: "the reduction of the uncertainties surrounding the operation enhances the usefulness of cost projections and increases the attention devoted to cost."[12]

It is conceivable, then that during this period, successful imitation may be realized in another country. As Hufbauer has observed, "Standardization aids and abets this migration in two ways -- longer production runs and proven production technology bring industry within the technical grasp of more nations; standardized goods are more easily marketed both because sales channels have been established and because feed back problems are less severe."[13]

In Figure 1.1, t_2 is the date of successful commercial production of the commodity by an imitating country; the time period t_0t_2 being termed the "imitation lag". Thus successful imitation results in a progressive decline in exports from the innovating to the imitating country. In the simplest version of Vernon's model, at some time, t_3, the external trade flow reverses direction, as the imitating country now possesses the comparative advantage in the commodity and begins exporting it back to the innovating country. "The hypothesis suggested in this paper has the United States exporting. . . products in the early stages of their existence, and importing them later on."[14]

These successive stages of standardization continue to evolve until "at an advanced stage in the standardization of some products, the less-developed countries may offer competitive advantages as a production location."[15] This expectation derives from the assumption that mature commodities, being highly standardized, will have easily accessible international markets, with market shares being determined primarily on the basis of lowest price, in addition, to be candidates for export from less developed countries, their production functions must be such as to employ large (relative) inputs of labor.[16]

Evidence for such trade patterns is largely impressionistic. Apart from the Japanese and Taiwanese success in developing significant overseas markets for standardized manufactured products, "all one sees are a few provocative successes such as some sporadic sales of newsprint from Pakistan, the successful export of sewing machines from India, and so on."[17]

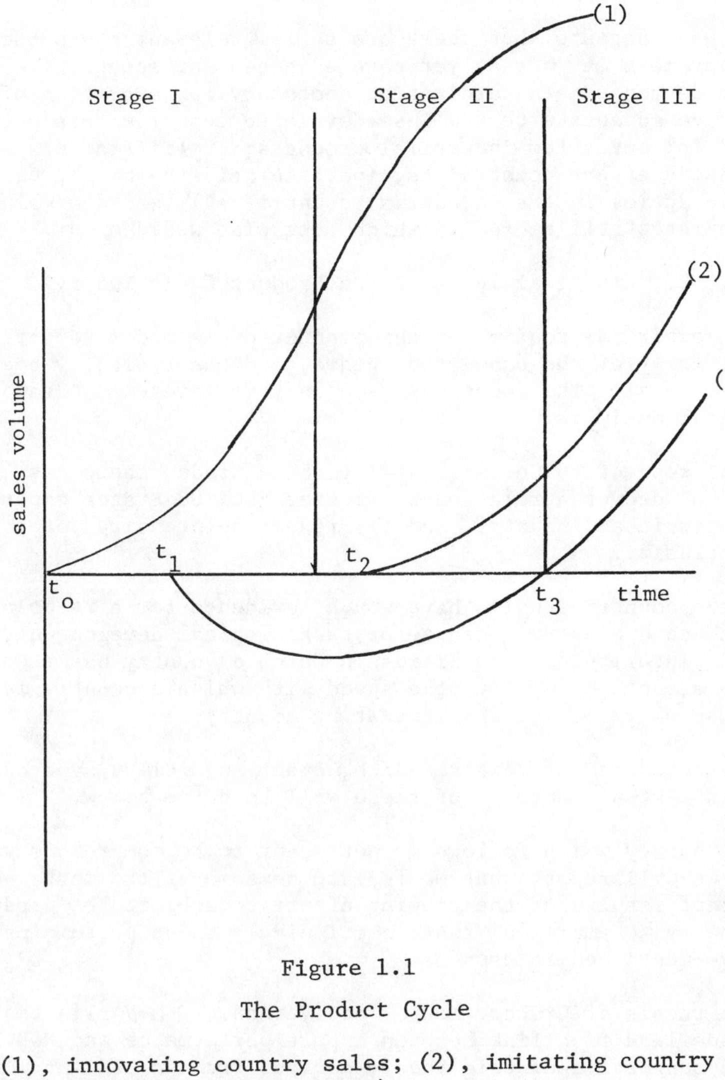

Figure 1.1

The Product Cycle

Path: (1), innovating country sales; (2), imitating country sales; (3), external (foreign trade).
Time period t_1t_3: foreign trade is comprised of exports from innovating to imitating country only; after t_3, the external trade flow reverses direction according to the shift in comparative advantage.

Vernon suggests that there are so few relevant cases because (i) the process may not as yet have advanced far enough, i.e., the required degree of standardization necessary for migration of comparative advantage to the less-developed countries has not obtained for but a few industrial products; or (ii) factors such as pervasive export constraints, (particularly the height of effective duties in the industrial countries[18]), and overvalued exchange rates stifle exports which otherwise would occur.[19]

2. Empirical Tests of the Product Cycle Theory

The empirical testing of the product cycle model to date has been primarily of the case-study genre, and has typically been cross-sectional. The focus has been on four inferences drawn from theoretical analysis.

With respect to United States foreign trade, there has been an attempt to uncover a relationship between United States export performance in an industry, and (i) research-intensity, or (ii) skill-intensity.

Inter-country studies have sought evidence for a relationship between a country's degree of technological development, and (iii) the maturity of commodities in which a country has a comparative advantage, or (iv) the speed with which a country is able to successfully imitate the innovating country.

In the course of this research, other subsidiary hypotheses have been tested. Several of these will be noted below.

The survey which follows is not meant to be comprehensive. Rather, we will report what we feel to be some of the most significant results of the growing effort to subject the product cycle theory to empirical test, particularly those pertaining to the above-mentioned inferences.

The thesis of Gruber, Mehta, and Vernon, (G-M-V), is that, "All roads lead to a link between export performance and R&D."[20] Evidence supportive of this thesis derives from an examination of the 1962 trade performance of nineteen United States (two-digit SIC) industries. In summarizing their results, the authors observe, "In sum, one derives a picture of high research effort being correlated with industries that experience substantial trade surpluses. . . It is in these industries that the United States trade advantage lies."[21]

Table 1.1

Research Intensity and World Trade Performance by United States' Industries, 1962
(Spearman Rank Correlation Coefficient for Indicated Cell)

Export Performance Measure	R_1	R_2
Total Trade of U.S. Industries		
E_1	0.69	0.74
E_2	0.79	0.69
Trade of U.S. Industries with Europe		
E_1	0.63	0.65
E_2	0.35*	0.48
Trade of U.S. Industries with non-Europe		
E_1	0.73	0.74
E_2	0.78	0.67
U.S. World Exports in 1962 as a Percentage of World Exports of:		
OECD countries	0.68	0.64
U.K.	0.28*	0.37
West Germany	0.08	0.24*
France	0.60	0.59

Source: G-M-V, op. cit., Table 3, Table 5.

E_1: Exports as a percentage of sales, 1962.

E_2: Excess of exports over imports as a percentage of sales, 1962.

R_1: Total R&D expenditures as a percentage of sales, 1962.

R_2: Scientists and engineers in R&D as a percentage of total industry employment, 1962.

* These coefficients are not significant at the 5 percent probability level.

G-M-V found that the five industries having the strongest research effort accounted for 39.1 percent of the sales of manufactured commodities, while accounting for 89.4 percent of United States total R&D expenditures, and 74.6 percent of company-financed R&D expenditures.[22]

In order to test further for a relationship between United States comparative advantage and research-intensity, several measures of United States comparative advantage were devised.[23] Spearman rank correlation coefficients were then calculated between ranking of the nineteen industries by each measure of comparative advantage, and a ranking by two measures of research intensity; total R&D expenditures as a percentage of 1962 sales, R_1, and scientists and engineers in R&D as a percentage of total industry employment, R_2. A summary of the results appears in Table 1.1.

These results generally lend support to the view that the international competitive strength of United States industries is directly related to their research intensity.

It is interesting to note that this relationship appears to be much stronger in the case of the United States trade with non-European as compared to that with European countries. This seems to suggest that Europe more efficiently challenges United States competitive strength in research-intensive industries.[24] That such is the case is supported by the results pertaining to U.K. and to German trade patterns as compared to that of the United States. "The result is that, suddenly almost all significant relationships disappear."[25] As the authors suggest, this is the expected result. These three countries are highly advanced, have high per capita incomes, and put a premium on innovation. As a result, their export profiles can be expected to be quite similar to that of the United States.

Keesing's thesis is virtually identical to that of G-M-V, "This article tests the hypothesis that R&D activity is associated with American competitive ability in manufacturing industries. . . "[26] 1962 U.S. exports in each of eighteen industries as a percent of the total 1962 Group of Ten[27] exports served as his measure of American competitiveness. Technical personnel as a percent of total value added, R_3, and technical personnel as a percent of total industry employment, R_2, served as measures of research intensity. Keesing results are reproduced (in part), in Table 1.2.

Table 1.2

Research Intensity and U.S. Exports Normalized by Group of Ten Exports, 1962.
(Linear and Spearman Correlation Coefficients for Indicated Cell)

1962 U.S. Exports as a % of Group of Ten Exports*

	Linear	Spearman
R_2	.88	.94
R_3	.66	.78

Source: Keesing, _op_. _cit_., p. 39.

R_2: Technical personnel as a percent of total industry employment

R_3: Technical personnel as a percent of total value added.

* All coefficients are significant at the .01 level of probability.

As Keesing observed,

> the correlation is much too high to be attributed to chance. . . there turns out to be a powerful relation between the intensity of R&D activity in American industries and their export performance. . .
>
> R&D 'explain' competitive trade success in manufacturing industries considerably better than any other variable tested. This finding is consistent with a view that the world economic role of the United States involved the systematic export of new products.[28]

Before leaving the G-M-V and Keesing papers, one additional finding should be noted. Orthodox international trade theory stipulates that a relatively capital-abundant country will enjoy a comparative advantage in capital-intensive commodities. The Leontief findings notwithstanding, since the U.S. is relatively capital-abundant, it will, according to orthodoxy, export capital-intensive commodities. The question then is, are new, research-intensive commodities capital-intensive? If so, then is it not capital-intensity, rather than research-intensity, which explains United States trade performance?

Both G-M-V and Keesing address themselves to this question. G-M-V reply that "two fairly sensitive measures of capital-intensity fail to display any systematic relation with research effort."[29] The results upon which this conclusion is based are reproduced in the first two lines of Table 1.3.

Keesing, on the other hand, finds a strong negative correlatio between capital requirements and trade performance. The Spearman rank correlation between an industry's capital requirements and its competitive trade performance is -0.78.[30] He also finds a negative relationship between research intensity and capital requirements, viz., the last line of Table 1.3.

Based on the evidence presented by G-M-V and Keesing on this score, it is very tempting to argue that successful United States trade performance in an industry is not due to exploitation of a relative abundance of capital, but rather to successful exploitation of a relative abundance of innovative capability.

An hypothesis consistent with the predications of the product cycle model is that the "relative maturity of the industry in which different countries possess competitive strength is negatively correlated with the stage of economic development reached by them. . . "[31]

Hirsch uses this hypothesis as the basis of an empirical test of the product cycle model. He begins from the premise that "it must be shown that the competitive position of the U.S., the world's foremost leader, is strongest in comparatively new electronic products, while being relatively weak in the more mature ones."[32]

In order to test the proposition that they are negatively correlated, individual sectors of the U.S. electronics industry were first ranked by their international performance, and then ranked on the basis of their relative maturity. Hirsch argues that the best guage of relative international performance might be the trends exhibited by exports, imports, and trade balances over a period of time, (1960 through 1962).[33]

Table 1.3

Relationship Between Measures of Research Intensity and Other Characteristics in United States Industries (Spearman Rank Correlation Coefficients for Indicated Cell)

Industry Characteristics	R_1(1962)	R_2(1962)
Depreciation expenses as a percentage of value added, 1957	-0.11*	0.03*
Net fixed assets as a percentage of value added, 1957	-0.09*	0.09*
Hickman net capital-output ratio, 1960	----	0.58[a]

Source: G-M-V, _op. cit._, Table 6; Keesing, _op. cit._, p. 42.

[a] R_2 is as of January, 1961.

* These coefficients are not statistically significant at the 5 percent level.

The product cycle model suggests that the rate of growth in output of an industry is inversely correlated with its maturity, viz., path (1), Figure 1.1. Early stages of production will be characterized by significant growth in output, with the latter becoming sluggish as the commodity attains maturity. Hirsch argues that it is consistent with this premise that value added in an industry should be inversely related to industry maturity, and therefore proposes "to guage the maturity of the different sectors of the U.S. electronics industry by comparing the rates at which their respective value added changed over the fifteen-year period 1947-62."[34]

Hirsch concludes that there appears to be a negative correlation between maturity and international competitiveness in the U.S. electronics industry.[35] As a check on Hirsch's conclusion, we calculated Spearman rank correlation coefficients between rankings of the sectors according to maturity and competitiveness. Table 1.4 presents growth rates in the trade balance of five sectors of the U.S. electronics industry for the period 1960-63, and data on the percent change in the index ratio,[36] for each sector, between 1947 and 1962.

When ranked by these two measures, there obtains a -.60 rank correlation coefficient between the international competitiveness of the U.S. electronics industry and its maturity. We therefore concur with Hirsch that "(t)hese findings are consistent with the predictions of the product cycle model."[37]

Table 1.4

Growth Rates in U.S. Trade Balance,
Selected Electronic Product Groups 1960-1963

Sector	Trade Balance % change 1960-63[a]	Index Ratio % change 1957-62[a]
Govt'l and Ind'l Products	57.0 (2)	171.0 (5)
Special-Purpose Tubes	11.0 (3)	87.0 (4)
Components & Accessories	77.0 (1)	23.0 (3)
Consumer Products	-284.0 (5)	-71.0 (1)
Receiving Tubes	-11.8 (4)	-51.0 (2)

Source: Table IV (2) and Table IV (4) in _Ibid_., pp. 70, 73.

[a]The number in parentheses following each growth rate is the rank of the sector.

Consideration of the skill-intensity of production in the various sectors of the electronics industry provides further evidence for consistency between the experience in the U.S. electronics industry and the predictions of the product cycle model. Table 1.5 presents non-production workers as a percent of production workers in each of five sectors in the electronics industry in 1958 and 1962. In both cases the Spearman rank correlation coefficient between the ranking of sectors by maturity and that by skill-intensity is -.90.

Table 1.5

U.S. Electronics Industry: Non-Production Workers as a percent of Production Workers, 1958, 1962

Sector	1958	1962	Rank
Government and Industrial	39.7	47.7	1
Special-Purpose Tubes	27.9	38.7	2
Components and Accessories	22.2	28.7	3
Consumer Products	21.8	20.4	4
Receiving Tubes	16.3	19.5	5

Source: *Ibid.*, Table IV (6), p. 80.

The product cycle model has also been tested by G. C. Hufbauer.[38] His is the most comprehensive test of the product cycle theory to date. Three rival accounts of international commerce are applied to the experience of synthetic materials: (i) the factor proportions account which views trade flows as a function of international factor price ratio differences; (ii) the scale economy account which depends upon increasing returns to scale and trade barriers; and (iii) the technological gap (or product cycle) account.

Hufbauer's major finding is that the factor proportions account is unable to predict international trade in synthetic materials. But both the scale economy theory and the technological gap account together explain most synthetic materials commerce.[39] A discussion of several of his other findings follows.

The product cycle model suggests that the more advanced nations will be faster imitators than their less advanced counterparts. That is, "(n)ew products spread very much like new fashions: from highly developed centres to somewhat less developed regions, then to remote areas."[40]

There is evidence suggesting that imitation in the synthetic materials field has progressed in just this fashion. Figure 1.2 illustrates this: 1957 aggregate imitation lags[41] are plotted against 1957 per capita production of manufactures for several countries. There is definitely a negative relationship between these two variables. Hufbauer gives the following example of how this process has in fact worked in this industry: ". . . America produced nylon fibre in 1938, and both Britain and France were manufacturing nylon in 1941. However, Greece did not manufacture nylon until 1961; and in Yugoslavia, this fibre is still 'under consideration'."[42]

Hufbauer introduces the idea that according to the product cycle account, initial external trade will be technological gap trade -- that flowing from the innovating country to potential imitating countries. As the product matures, and there is imitation, low-wage trade will increasingly supplant technological gap trade, as imitating countries are increasingly able to exploit their relative abundance of unskilled labor. This phenomenon is illustrated by path (3) in Figure 1.1.

If the experience of the synthetic materials industry is consistent with the predictions of the product cycle model, as each commodity in the industry ages, an increasing proportion of total external trade in that commodity should be accounted for by low-wage trade. Hufbauer finds such to be approximately the case. He concludes that "(w)hile no exact relationship exists between the two variables, it does appear that low-wage trade has a greater importance in an older group."[43]

Lastly, a basic theme of the product cycle literature is that since technology spreads from very developed to less developed areas, advanced countries should tend to specialize in the export of newer commodities, in this case, newer synthetic materials. Hufbauer offers evidence suggesting that advanced nations export a larger fraction of their production of new synthetics than of old synthetics.[44]

In the case of plastics and synthetic rubber exports, the United States "exhibits a pronounced tendency to specialize in her newer plastics and synthetic rubbers. In fact the tendency is so strong that a logarithmic relationship apparently best suits the data."[45]

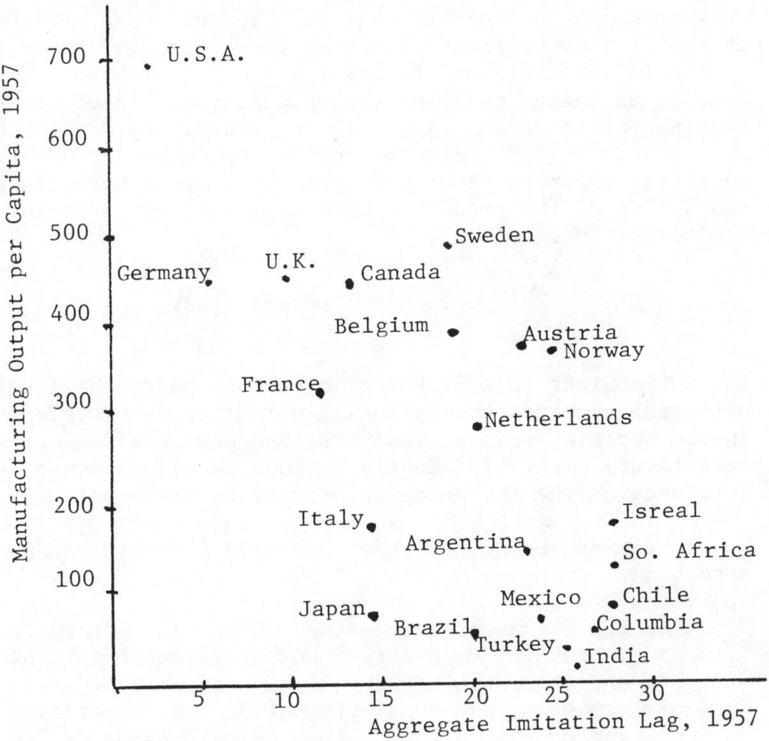

Figure 1.2

Imitation Lag and Per Capita Output of Manufactures

Source: G.C. Hufbauer, _Synthetic Materials & the Theory of International Trade_, Diagram 5-2.

American man-made fiber exports follow almost an identical pattern. "Clearly, America specializes in exporting the newer nylon, polyester, and acrylic fibres and the newer forms of viscose rayon, rather than the older man-made fibres."[46]

The preceding comprises a brief survey of the highlights of what we consider to be the most significant empirical tests of the product cycle hypothesis to date. Taken as a whole, it is very tempting to conclude that the product cycle hypothesis is as empirically "correct" as it is theoretically interesting. But as we are cautioned by G-M-V, "A chain of hypotheses as complex as these (those of the product cycle) needs extensive testing before it can gain much in credibility."[47] Much further research is needed, both theoretical and empirical. It is our hope that the results of our study will contribute to the tenability of the product cycle theory, albeit only for a few additional industries.

3. The Present Study

Set against this backdrop, our study addresses itself to obtaining a better understanding of the product cycle hypothesis in the hopes of gaining insights into how received comparative cost theory might efficiently include consideration of the influence of the technological factor on trade patterns.

In commenting on the above-surveyed literature, Johnson has observed:

> The fundamental question, which has been raised in general by a large body of recent empirical and theoretical work on the economics of domestic industrial production and competition and on the role of technological change in economic growth, and which has been raised in particular by the recent empirical work on the technological factor in international trade is whether orthodox international trade theory has once again reached a position of commitment to an antiquated and misleading model of the production process, which has to be reflected to further the progress of economic science.[48]

Our goal is to demonstrate both the richness of Vernon's approach and its ability to allay Johnson's fears. Our method involves the construction of general equilibrium models of dynamic comparative advantage. These models include Vernon's assumptions and are capable of explaining the results to which Johnson alludes.

In order to simplify the analysis, the product cycle literature to date assumes that new commodities are homogeneous. If this were correct, all external trade in new commodities would be unilateral -- see Figure 1.1 -- with "low-wage" trade replacing "technological gap" trade. But bilateral and multilaterial trade in new commodities is common, viz., computers, transistor circuitry, and optical equipment. This suggests that new commodities are often non-homogeneous. That such non-homogeneity as a characteristic of new commodities does not destroy the product cycle theory is one major goal of our study. We will demonstrate that the Vernon model is easily amended to include non-homogeneous commodities.

In pursuing this goal, we build two general equilibrium models of dynamic comparative advantage: one each for homogeneous and non-homogeneous new commodities. The distinguishing characteristic of these models is "endogenized" technological change of the type envisioned by Vernon, Posner, Hufbauer, et. al.

Our initial problem in this effort was the determination of how best to "endogenize" (in the sense of Chipman[49]) technological change. We were greatly influenced in this choice by Chipman and Arrow. Chipman suggests that there are two ways to approach this problem. The traditional, static, neo-classical production function -- that to which Johnson refers above -- can be dynamized by providing a specific rule for changing it over time. Alternatively, the static production function can be discarded -- as Johnson seems to suggest is necessary -- and a specific dynamic function can be defined in its stead.[50]

For our purposes, Chipman's first method of "endogenization" was chosen; we agreed with Chipman's argument that there is "a distinct advantage in retaining the static production function concept . . . (in that) the modern Hechscher-Ohlin-Lerner-Samuelson theory of international trade, with given or exogenously changing production functions has been fairly worked out."[51]

Having made this choice, there remained the problem of providing the rule with which we were to dynamise the static, neoclassical production function. As a source for this rule, we appealed to Arrow. In commenting on Chipman's paper, cited above, Professor Arrow observes that

> Professor Chipman minimizes the importance of learning by doing as a cause of technological progress. But the notion of the 'product cycle', so much emphasized at this conference, is certainly an example of learning by doing; it is assumed that requirements for certain skill factors decline with experience in production.[52]

Provided with the modus operandi, we proceeded to examine international trade and technology within models of dynamic comparative advantage.

Our study begins in Chapter Two with the presentation of a simple, two-country, three-factor, multicommodity model of dynamic comparative advantage, given the innovation of one new homogeneous commodity. With this model, we derive the necessary and sufficient conditions which must be satisfied if the central assertions of the product cycle model are to obtain; in particular, that innovation obtains in the lion's share of cases in the technologically most developed economy; followed, via successive stages of standardization in manufacturing processes, by migration of comparative advantage and production in the direction of technologically less developed economies.

In the course of this endeavor, we essay to illumine facets of Vernon's theory which are implicit in his formulation. In particular, we investigate the nature of the time profile of the market price of a new commodity. The duration of imitation lags is also examined. Vernon hints that the duration of imitation lags across an array of potential imitators will be affected by the nature of those imitators. We investigate the aspects of economies likely to have an impact on imitation lag duration, seeking to ascertain both the direction and magnitude of same.

Vernon innovates, and Hufbauer and Hirsch[53] echo the idea of a "pecking-order" which suggests that a ranking of economies by (declining) relative (to the innovating country's) technological development also ranks them by imitation lags of increasing length. We make an effort to discern the necessary and sufficient conditions which must be satisfied if a "pecking-order" is to obtain.

An integral part of an analysis of international trade in new commodities is a consideration of the issue of consumer demand. In particular, an effort should be made to construct a model which will describe the growth in aggregate demand for new commodities. This is the goal of Chapter Three. We incorporate Lancaster's "new approach" to consumer theory,[54] and the concept of consumer learning into dynamic microeconomic models of consumer behavior in the ith new commodity. Aggregation over the resulting J microfunctions and consideration of the impacts of learning, income, and substitution effects, as well as certainty and uncertainty, results in models describing the equilibrium time path of quantities demanded for both homogeneous and non-homogeneous new commodities.

The analysis of Chapter Three serves as the cornerstone for the product cycle model of a non-homogeneous commodity presented in Chapter Four. After indicating that the Posner-Vernon-Hufbauer analysis cited above rests on an (implicit) assumption that all new commodities are homogeneous, we construct and test, via simulation, a bilateral trade product cycle model of the ith new commodity, when the latter is assumed non-homogeneous.

Chapter Five presents the findings of our statistical analysis. In an effort to determine the extent to which the facts of observation refute the assumptions and conclusions of our theoretical analysis, we subject several hypotheses derived therefrom to empirical analysis. Our general conclusion based on these tests is that for our sample of commodities over the period of observation, observed deviations of the facts from the theory are sufficiently miniscule for us to remain optimistic that the approach ennunciated by Vernon will prove to successfully incorporate the technological factor into received comparative cost models.

FOOTNOTES

[1] Harry G. Johnson, "Comparative Cost and Commercial Policy Theory for a Developing World Economy," Wicksell Lectures 1968, p. 12.

[2] C.P. Kindleberger, The Dollar Shortage, Irving B. Kravis, "Availability and Other Influences on the Commodity Composition of Trade," Journal of Political Economy, April, 1956 Erik Hoffmeyer, Dollar Shortage.

[3] Kindleberger, op. cit., pp. 12-17.

[4] Hoffmeyer, op. cit., Research intensity of traded goods was determined on the basis of data on company spending on R&D as a percent of sales, and patents granted since 1900; pp. 161-162.

[5] Ibid., p. 160.

[6] Ibid., p. 159.

[7] M.V. Posner, "International Trade and Technical Change," Oxford Economic Papers, pp. 322-341.

[8] Ibid., p. 322.

[9] Vernon, R., "International Investment and International Trade in the Product Cycle," Quarterly Journal of Economics, 1966, pp. 190-207.

[10] Ibid., p. 192

[11] Ibid., p. 195.

[12] Ibid., p. 196.

[13] G.C. Hufbauer, "The Impact of National Characteristics and Technology on the Commodity Composition of Trade in Manufactured Goods," in R. Vernon, ed., op. cit., p. 189.

[14] Vernon, op. cit., p. 201.

[15] Ibid., p. 202.

[16] Ibid., p. 203

[17] Ibid., p. 205.

[18] cf., Bela Balassa, Trade Liberalization Among Industrial Countries, Objectives and Alternatives, Chapter Three.

[19]Vernon, *op. cit.*, p. 205.

[20]Gruber, W., Mehta, D., and Vernon, R., "The R&D Factor in International Trade and International Investment of United States Industries," *Journal of Political Economy*, Feb., 1967.

[21]*Ibid.*, p. 30. Gruber and Vernon have elsewhere examined virtually the same thesis with a sample of 1964 trade of ten trading areas and twenty-four industries. They conclude that never-mind the measure of research-intensity chosen, "(t)echnology-intensive industries occupy an especially prominent place in the export mix of the United States and a less-then-proportionate place in the export mix of most areas toward the lower end of the development spectrum." Gruber and Vernon, "The Technology Factor in a World Trade Matrix," in R. Vernon, ed., *The Technology Factor in International Trade*, p. 244.

[22]G-M-V, *op. cit.*, p. 25.

[23]See Table 1.1.

[24]*Ibid.*, p. 27.

[25]*Ibid.*, p. 26.

[26]D.B. Keesing, "The Impact of Research and Development on United States Trade," *Journal of Political Economy*, February, 1967.

[27]United States, United Kingdom, Federal Republic of Germany, France, Italy, Belgium, Netherlands, Sweden, Canada, Japan.

[28]Keesing, *op. cit.*, pp. 39, 45.

[29]G-M-V, *op. cit.*, p. 28.

[30]Keesing, *op. cit.*, p. 42.

[31]Seev Hirsch, *Location of Industry and International Competitiveness*, p. 62.

[32]*Ibid.*, pp. 63, 64.

[33]*Ibid.*, p. 71.

[34]*Ibid.*, p. 73.

[35]*Ibid.*, p. 75.

[36]The index ratio is calculated by dividing the index of value added of a sector by that for the total industry for each year between 1947 and 1962.

[37] *Ibid.*, p. 75.

[38] G. C. Hufbauer, *Synthetic Materials and the Theory of International Trade.*

[39] *Ibid.*, p. 110.

[40] *Ibid.*, p. 91.

[41] The aggregate imitation lag is a weighted average of national imitation lags in individual products in the industry. The weight in each case is the product's share of current world exports of that commodity. cf., *Ibid.*, pp. 82, 83.

[42] *Ibid.*, p. 92.

[43] *Ibid.*, p. 95.

[44] *Ibid.*, p. 95.

[45] *Ibid.*, p. 105.

[46] *Ibid.*

[47] G-M-V, *op. cit.*, p. 22.

[48] H.G. Johnson, "Theory in Relation to Empirical Analysis," in Vernon, ed., *op. cit.*, p. 11.

[49] J.S. Chipman, "Induced Technical Change and Trade Patterns," in Vernon, ed., *op. cit.*, pp. 95-127.

[50] *Ibid.*

[51] *Ibid.*, pp. 96, 97.

[52] Arrow, K.J., in commenting on *Ibid.*, p. 131.

[53] Hufbauer, *op. cit.*; Hirsch, *op. cit.*

[54] K. Lancaster, "Change and Innovation in the Technology of Consumption," *American Economic Review, Papers and Proceedings*, May, 1966, pp. 14-23; _________, "A New Approach to Consumer Theory, *Journal of Political Economy*, April, 1960.

Two

A Product Cycle Model for a Homogeneous Commodity[1]

1. Introduction

The product cycle model postulates that in the production of a new commodity, the input mix which is optimal at early stages of production differs in a systematic fashion from that which is optimal at later stages of production; or what is the same thing, "there is a systematic bias in the change in the production function required to produce the commodity."[2] During early stages of production of a new commodity, commercially feasible processing and manufacturing methods must be developed, thereby necessitating a relatively large input of factors embodying special skills. As this development proceeds, there is associated with it an increasing routinization of production technique which acts to progressively reduce the requirement (relative to unskilled inputs) for skill-embodied inputs, i.e., the skill-output ratio at constant factor prices declines over time.

Vernon asserts that while the technological most sophisticated countries have relatively high labor costs, this disadvantage may be outweighed in the case of new commodities (at least in early stages of production) by a ready accessibility to a variety of factors embodying special skills.[3] That is, since the technologically most developed country will be abundantly endowed relative to less developed countries in these skill-embodied factors, it will realize a comparative advantage in commodities whose production is intensive in such factors: new commodities at early stages of production. But as the product cycle unfolds, the above-mentioned systematic variations in optimal factor proportions come into play. As the product becomes standardized, thereby imparting a relatively greater role to unskilled factors, shifts in the locus of comparative advantage (and thereby production) in the direction of technologically less developed countries obtain.

As Jones points out, Vernon's assertions can be interpreted within the framework of a three factor model: capital, ordinary labor, and a third factor which can be termed skill.[4] Hence the goal of this chapter.

We propose to offer a formal statement of Vernon's model couched within the framework of a variation on the standard H.O.S. model. We defend this as a valuable exercise on its own merits. It demonstrates that Vernon's assertions are consistent with a more sophisticated form of the traditional (static) theory of comparative costs, and it brings into focus several significant facets of Vernon's contribution. In particular, we note necessary and sufficient conditions to be satisfied if a new commodity is first to be produced in the technologically most developed country, and those to be met if a switch in comparative advantage is to obtain.

The chapter begins, in Section 2, with the list of assumptions which will hold throughout. In Section 3, a two-country, single-new-commodity, three factor model of dynamic comparative advantage is presented. Systematic variation over time in optimal factor proportions is introduced, reflecting the dynamic bias in required input mix which characterizes the product cycle.

In Section 4, necessary and sufficient conditions for existence and uniqueness are offered. That is, there exists a point in (finite) time, T*, such that the locus of minimum production cost (and therefore comparative advantage) shifts from country 1 to country 2; and secondly, that T* is unique. In Section 5, the impact on T* of (i) different rates of chance in the required input mix, and (ii) changes in the level of technological sophistication of potential imitating countries are considered. This analysis generates a theorem which provides the necessary and sufficient conditions from a ranking of dates at which potential imitating countries will realize a pair-wise cost advantage vis à vis the innovating country to obtain.

2. Assumptions

1. There exist two countries, three factors of production (capital, ordinary labor, and a third factor to be defined as skill in the sense of Jones[5]);

2. each country is endowed with given amounts of (homogeneous) labor and skill (in differing ratios) whose supply is perfectly inelastic;

3. capital is homogeneous and mobile both within and between countries; thus there exists a world supply price, r;

4. labor and skill are immobile between countries, but mobile within them;

5. international factor price equalization is incomplete; differing (between countries) rates of return obtain for labor and skill within each country;

6. the conditions of pure and perfect competition are satisfied; market price is equated to marginal cost, transportation is costless, information is perfect, adjustment is instantaneous, and there is no uncertainty;

7. each country consists of N sectors of which the Nth is the single new commodity industry; the latter being so small vis à vis the N-1 industries, as not to affect national factor prices;

8. the new good is produced with a concave, linear homogeneous production function which is the same in each country;

9. the required input mix changes through time such that at constant factor prices, production becomes relatively more labor and relatively less skill-intensive.

Assumptions 3, 5, and 9 require amplification. Assumption 3 permits attention to be focused on the role played by the availability of this third factor, skill, as a cause of comparative advantage.[6]

Assumption 5 follows directly from the fact that labor and skill are assumed homogeneous and intranationally mobile, and that there is a single new sector (industry) in an economy of N industries. Since labor and skill are mobile and homogeneous, a single national rate of return obtains for each.

Assumption 9 is basic to our model. An explicit statement of Vernon's basic assertion, this modification of the H.O.S. model is necessary to admit a formalization of what we consider to be central aspects of the product cycle theory. We recognize that there remains the larger issue of the nature of a model which would have assumption 9 as one of its results, but this goes beyond the goal of this chapter.

3. The Model

Given this list of assumptions, we can develop a model capable of determining in which of the two countries first production will occur; and the point in time when the cost advantage and therefore production will shift to the second country.

The criterion for determining the locus of cost advantage will be minimum marginal cost. That is, country 1 will have a cost advantage for all t such that MC_{1t} is less than MC_{2t}; which is to say that for all such t, production of the commodity will take place in country 1 only; and conversely for all t such that $MC_{1t}>MC_{2t}$ obtains.

The commodity is produced by the production technology given by (2.1), where $A_1=A_2$, and a, b, and j, are the (partial) elasticities of output with respect to variations in capital (K), labor (L), and skill (S), respectively.[7] The cost equation for production of the commodity is given by (2.2). The relative magnitudes of the factor prices as between country 1 and country 2 and given by (2.3) and (2.4)[8]. These reflect the conditions set out above: country 1, being more developed than country 2, exhibits a lower skill return and higher labor return relative to that of internationally mobile capital, whose return is identical in each country.

$$Q = AK^aL^bS^j; \qquad a+b+j = 1 \tag{2.1}$$

$$C_i = w_iL_i + s_iS_i + rK_i; \qquad i = 1,2; \tag{2.2}$$

where (given perfect competition prevails):

$$w_i = p_iQ_{L_i}, \ s_i = p_iQ_{S_i}, \text{ and } r = p_iQ_{K_i}.$$ [9]

$s_1/r_1<s_2/r_2$, or $s_1/s_2<1$; which can be written:

$$V = \log s_1 - \log s_2<0. \tag{2.3}$$

$w_1/r_1>w_2/r_2$, or $w_1/w_2>1$; which can be written:

$$W = \log w_1 - \log w_2>0. \tag{2.4}$$

A: The scale efficiency coefficient, or the measure of total factor productivity;

s_1, s_2: rates of return to skill;

w_1, w_2: rates of return to labor.

The ith country's marginal cost of production as a function of the level of production and the elasticities of output with respect to the three factors can be determined by minimizing (2.2) subject to (2.1). Given the Lagrangian function:

$$U = AK^a L^b S^j + \lambda(C_i - w_i L_i - s_i S_i - rK_i);$$

the first order conditions for a constrained optimum are:

$$\partial U/\partial K = a(Q/K) - \lambda r = 0$$
$$\partial U/\partial S = j(Q/S) - \lambda s_i = 0$$
$$\partial U/\partial L = b(Q/L) - \lambda w_i = 0$$
$$\partial U/\partial \lambda = c_i - w_i L_i - s_i S_i - rK_i = 0$$

Given the first-order conditions, and recalling the assumption of perfect competition, we can write:

$$K_i = MC_i Qa/r, \; L_i = MC_i Qb/w_i, \text{ and } S_i = MC_i Qj/s_i; \qquad (2.5)$$

since $\lambda = 1/MC_i$.

Substituting these into (2.1) gives:

$$Q = A(MC_i Qa/r)^a (MC_i Qb/w_i)^b (MC_i Qj/s_i)^j;$$

and then solving for marginal cost, MC_i:

$$MC_i = A^{-1}(r^a w_i{}^b s_i{}^j)(a^a b^b j^j)^{-1};$$

or:

$$MC_i = \bar{M}(Ab^b j^j)^{-1} w_i{}^b s_i{}^j; \; \bar{M} = (r/a)^a, \; i = 1,2 \qquad (2.6)$$

Given assumption 5, the dynamic behavior of MC_i is determined solely by the initial values and the time profiles of the factor elasticities of output.

Let:

$$j_t = h(t) \text{ such that } dj_t/dt<0, \text{ and } d^2j_t/dt^2>0; \tag{2.7}$$
$$b_t = g(t) \text{ such that } db_t/dt>0, \text{ and } d^2b_t/dt^2<0;$$

where: $h(0)>g(0)$ reflecting the relative skill intensiveness of production of new commodities upon innovation. The capital elasticity is assumed constant[10], and production linear homogeneous. Therefore: $h(t) + g(t) = (1-a)$, a constant which will be defined as E: and:

$$db_t/dt = -dj_t/dt.$$

Given (2.7), (2.6) can be written:

$$\log MC_{it} = \log\bar{M} - b_t \log b_t - j_t \log j_t + b_t \log w_i + j_t \log s_i - \log A, \tag{2.9}$$
$$i = 1,2.$$

The question, "In which country will first production occur?" can now be answered. Consider $\log MC_{10} - \log MC_{20}$; if this difference is less (greater) than zero, first production will occur in the first (second) country. Letting $Z_t = \log MC_{1t} - \log MC_{2t}$: $Z_0 = b_0 \log w_1 + j_o \log s_1 - b_o \log w_2 - j_o \log s_2$; which from (2.3) and (2.4) can be written:

$$Z_0 = b_o W + j_o V. \tag{2.10}$$

A necessary and sufficient condition for country 1 to enjoy initial production of the new commodity is that Z_o be negative; which from (2.10) implies that $|j_o V|$ be greater than $|b_o W|$, or that

$$j_o > -b_o W(V)^{-1}. \tag{2.11a}$$

Recalling that $E = b_o + j_o$, the condition that Z_o be negative can be written:[11]

$$j_o > -EW(V-W)^{-1}. \tag{2.11b}$$

Since V-W is negative, and greater in absolute value than W, and E is less than one, $0<-EW(V-W)^{-1}<1$, and condition (2.11b) can be met without violating the initial condition that j_t be always positive and less than one.

Another transformation of the condition that Z_o be negative indicates the economic significance to be given to (2.11a) and (2.11b). It follows from (2.11a) that

$$j_o/b_o > -W/V, \tag{2.12}$$

when country i enjoys initial production of a new commodity. But from the first order conditions for a constrained optimum given above, it follows that the left-hand side of (2.12) is equal to $S_o s_i/L_o w_1$. Therefore (2.12) can be written:

$$S_o s_1/L_o w_1 > -W/V. \tag{2.12a}$$

The right-hand side of (2.12a) can be defined as the "relative labor disadvantage" of country i -- the extent to which her relatively rationed labor supply is offset by a relatively abundant skill endowment. The left-hand side of (2.12a) measures the factor share of skill relative to that of labor when producing with optimal input mix in country 1. Thus, the condition which must be satisfied if country 1 is to enjoy initial production is that the relative share of the skill factor must exceed the relative labor disadvantage of country 1. In the event that for the commodity in question this condition is not satisfied, the locus of first production would not be country 1. Therefore it is not inevitable that the locus of initial production of a new commodity will be the technologically most developed economy.

4. The Switching Point: Existence and Uniqueness

Given that the necessary and sufficient conditions are satisfied and initial production occurs in country 1, the question of switching must be considered. What are the necessary and sufficient conditions for a shift in the location of production cost advantage to country 2?

Necessary Conditions. Taken as given that (2.7) holds for all t, the necessary conditions for a switch are:

i) At some point in time, T*:

$$Z_{T*} = \log MC_{1T*} - \log MC_{2T*} = 0.$$

From (2.10) it follows that $b_{T*W} + j_{T*}V$ must equal zero; or

$$j_{T*} = -EW(V-W)^{-1}. \tag{2.13}$$

ii) $Z_t \overset{<}{\underset{>}{=}} 0 \forall t \overset{<}{\underset{>}{=}} T^*$. From (2.13) it follows

$E = -j_{T^*}(V-W)(W)^{-1}$; and substituting in for E in

$Z_t = EW + j_t(V-W)$, gives:

$$Z_t = (j_t - j_{T^*})(V-W). \qquad (2.14$$

Observe that $V-W<0 \forall t$, and since $dj_t/dt<0$:

$j_t \overset{>}{\underset{<}{=}} j_{T^*} \Leftrightarrow t \overset{<}{\underset{>}{=}} T^*$. Therefore for all $0<t \overset{<}{\underset{>}{=}} T^*$,

$(j_t - J_{T^*}) \overset{>}{\underset{<}{=}} 0$ obtains which together with $V-W<0$, implies

$Z_t \overset{<}{\underset{>}{=}} 0 \; \forall t \overset{<}{\underset{>}{=}} T^*$.

iii) A final necessary condition obtains in instances where the scale efficiency coefficient is constant over time.[12] The elasticity of the isoquant at the optimal factor mix for the given factor prices must be less than one.[13]

Sufficient Conditions. The sufficient conditions for a switch follow from a consideration of the shape taken by the time path of the marginal cost functions under different assumptions regarding the behavior of the scale efficiency coefficient, A, over time.

Total factor productivity (A) assumed invarient over time. At any moment in time, a producer (or potential producer) has available to him an array of technologies with which a given commodity could be manufactured. But only the efficient technology will be chosen: that whose optimal factor mix minimizes the total cost of a given level of production and set of factor prices. A shift from an existing technology to a new one requires that the latter (the "marginal" technology) be at least as efficient as the former, which is to say that the marginal cost of producing with the new technology (and its associated optimal factor mix) must be less than or equal to that of producing with the current technology.

The impact of the efficiency criterion is that the above-prescribed variations in elasticities of putput with respect to changes in labor and skill, while necessary for variations in optimal input mix, may not be sufficient. Differentiating (2.9) with respect to time -- remembering the assumed constancy of the total factor productivity coefficient -- gives:

$$d\log MC_{it}/dt = dj_t/dt(\log b_t - \log w_i + \log s_i - \log j_t);^{14} \qquad (2.15)$$

which on the assumption of profit maximizing firm behavior can be written:

$$d\log MC_{it}/dt = dj_b/dt(\log L_t - \log S_t).^{15} \qquad (2.16)$$

From (2.15) it follows that:

$$d\log MC_{it}/dt \gtreqless 0 \Longleftrightarrow \log L_t \lesseqgtr \log S_t;$$

$$\text{or: } L_t/S_t \lesseqgtr 1.$$

But as shown in Figure A2.1 (See Appendix 2), there exists a direct relationship between the labor-skill elasticity of the isoquant at the optimal factor mix (for given factor prices), N_{LS}, and the labor-skill input mix itself. It can be seen from Figure A2.1 that the following relationship holds:

$$N_{SL} \gtreqless 1 \Longleftrightarrow L_t/S_t \lesseqgtr 1 \Longleftrightarrow \log L_t \lesseqgtr \log S_t:$$

and therefore:

$$d\log MC_{it}/dt \gtreqless 0 \Longleftrightarrow N_{LS} \gtreqless 1 \qquad (2.17)$$

Therefore with A constant, for all t such that $N_{LS} \gtrless 1$ obtains, the only efficient technology is the initial technology, and no variation in optimal factor mix obtains. It follows that a necessary condition for emerging technologies to be adopted within this context is for $N_{LS}<1$ to obtain. But this is not sufficient; in Figure 2.1, N_{LS}<obtains over the T^1T^2, but none of these technologies will be adopted. Thus the sufficient condition for variations in optimal factor mix to obtain at time t, under the assumption of constant total factor productivity, is: $\log MC_{it} < \log MC_{i0}$.

Only in the case where $N_{LS}^{o}<1$ obtains with the initial technology will marginal cost be a monotone decreasing function of time, thereby guaranteeing that the marginal technology is superior to the existing one, thus producing smooth variation in optimal factor mix over time.[16] In any other situation,

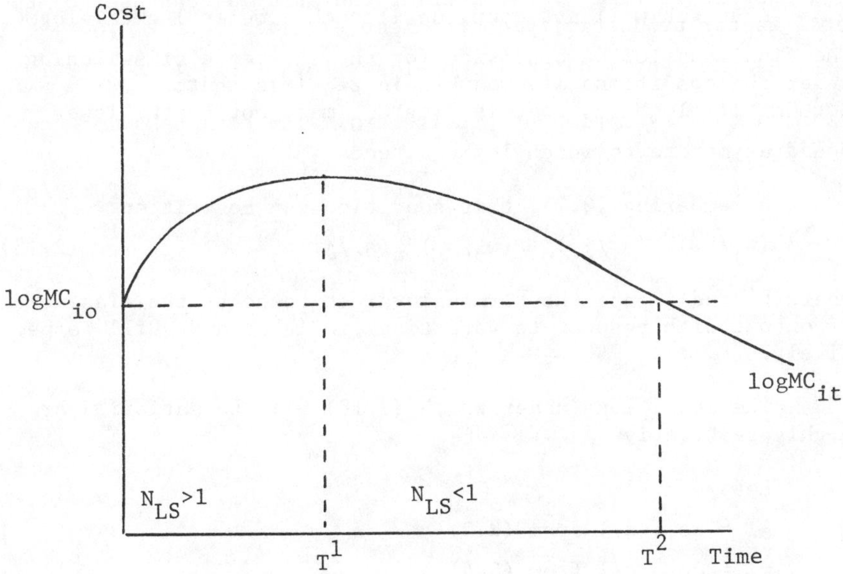

Figure 2.1

The Time Profile of $LogMC_{it}$

discontinuous shifts in optimal factor mix would obtain: a constant technology for an initial period followed by continuous variations in optimal input mix.

This discontinuity has implications for the conditions which must be satisfied if a shift in the location of cost advantage is to obtain. The situation is depicted in Figure 2.2, (see page 38), when $N_{LS}^{0}>1$ and A is constant.

Country 1 innovates the commodity with a marginal cost of $\log MC_{10}$. Over time new technologies become available, but they fail to satisfy the efficiency criterion and therefore are not adopted. Conditions (2.13) and 2.14) obtain at T*, but the minimum production cost site will not shift to country 2 because the initial technology is still the only efficient one at this juncture.

A switch will not occur until such time as $\log MC_{10} = \log MC_{2T}$. The final sufficient condition for the existence of switching under the conditions assumed herein requires that: $\log MC_{10} > \lim \log MC_{2T}$, that is, $\log MC_{10}$ must not be the lower bound asymptote to which $\log MC_{2T}$ tends.

Remembering (2.7), this condition can be written:

$$(w_1/b^o)^{b^o}(s_1/j^o)^{j^o} > (w_2/b^T)^{b^T}(s_2/jT)^{j^T} \qquad (2.18)$$

where b^T and j^T are the terminal values taken by the elasticities of output with respect to variations in labor and skill respectively.

The conditions under which (2.18) will be satisfied are highly restrictive.

Let:

$s_2 = zs_1$, $z>1$ from (2.3)

$w_2 = cw_1$, $c<1$ from (2.4)

$j^o = ej^T$, $e>1$ from (2.7).

Now if, and only if, the following special case obtains: $b^o=j^T$ and $j^o=b^T$; (2.18) can be rewritten: $w_1/w_2^e \; s_2/s_1^e$; and substituting for s_1 and w_1:

$$(w_2/s_2)^{(1-e)} > cz^e \Longrightarrow w_2/s_2 > c^{(1/1-e)}z^{(e/1-e)}. \qquad (2.19)$$

Only in the event that the factor price ratio in country 1 satisfied (2.19) will all sufficient conditions for switching under current assumptions be satisfied.

Total factor productivity assumed to be a monotone increasing function of time. If we now suspend the assumption of a constant technical efficiency coefficient, we admit technological progress in addition to technological change. This allows us to derive the sufficient condition for a monotone decreasing marginal cost function.

Let:

$A = A(t)$ such that $dA/dt>0$, $d^2A/dt^2<0$; then

$$d\log MC_{it}/dt = dj_t/dt(\log L_t - \log S_t) - (dA/dt)A^{-1}; \qquad (2.20)$$

and

$$dj_t/dt(\log L_t - \log S_t) < dA/dtA_t^{-1}; \quad dA/dtA_t^{-1} > 0 \forall t \qquad (2.21)$$

is sufficient to guarantee that the marginal cost function is a monotone decreasing function of time.

Satisfaction of (2.20) guarantees the existence of continually emerging technologies, and given that the necessary conditions (2.13) and (2.14) are met, that a switching point exists. In fact, under conditions of monotone decreasing marginal cost functions, the switching point will be T*, the point at which satisfaction of (2.13) and (2.14) obtains.

<u>Uniqueness</u>. It can be shown that if V and W are constant over time, the point $t = T^* \ni Z_{T^*} = 0$ is unique.

It must be shown that $dZ_t/dt > 0 \forall t > 0$. From (2.13),

$$dZ_t/dt = dj_t/dt(V-W)$$

Since $dj_t/dt < 0\ \forall t > 0$, and (V-W) is negative, then

$$dZ_t/dt > 0\ \forall t > 0.$$

Summarizing, the necessary and sufficient conditions for existence are:

1. $j_{T^*} = -EW(V-W)^{-1}$; s.t. $j_{T^*} + b_{T^*} = E$;
2. $Z_t < 0\ \forall\ 0 < t < T^*$;
3. $Z_t > 0\ \forall\ t > T^*$;
4. $dj/dt\ (\log L_t - \log S_t) < A'(t)A_t^{-1}$.

Given that these conditions are satisfied, T* will exist, and will be unique. This result is shown graphically in Figure 2.3. Z_t is drawn to be concave from below. That this is correct can be seen by differentiating (2.20) with respect to time.

$$\frac{d^2 Z_t}{dt^2} = \frac{d^2 j_t}{dt^2}(V-W) \qquad (2.22)$$

which, since $\frac{d^2 j_t}{dt^2}$ is positive by (2.7), is negative for all positive t.

5. The Response of T* to Variation in Selected Parameters

Having presented necessary and sufficient conditions for existence and uniqueness of the switching point, we now turned to the behavior of T* when selected parameters of the model are allowed to vary.[17]

The velocity of the product cycle. Consider first the impact on T* should the rate at which the product cycle proceeds vary. In other words, what is the relationship between T_1^* and T_2^* if, *ceteris paribus*, the rate of change in required factor mix in the first instance (T_1^*) is greater than in the second (T_2^*)?

Since dj_t/dt is equal to $-db_t/dt$, the magnitude of the former determines the rate of change in required factor mix at constant factor prices. In (2.13), it was seen that T* is that point in time when the skill-output elasticity is equal to $-EW(V-W)^{-1} = \bar{J}$. At issue, then, is how the length of time it takes j_t to take on the critical value, $\bar{J}$, varies when the rate of change of the former varies. The solution is straightforward as can be seen by examining Figure 2.4. Beginning at an initial value of j_o, the skill-output elasticity can follow three alternative routes, $h_c(t)$, c = 1,2,3. The rate of change required factor mix at any point in time declines proceeding from path h_1 to path h_3; which is to say $\underline{h_1'} > h_2' > h_3'$, evaluate at any t. The intersection of $h_c(t)$ and $\bar{J}\bar{J}$ determines T_c^*. It is clear that T* varies inversely with the rate of change in required factor mix.

The conclusion to be drawn from this result is clear: other things being equal, the duration of country 1's cost advantage in the nth new commodity vis à vis any potential immating country will be of greater (lesser) length, the slower (faster) the rate at which the product cycle proceeds. By way of example, should the United States as the technologically most developed economy undertake initial production of a new commodity, the length of the period during which she can successfully exploit her relative skill advantage vis à vis any potential imitating country will be inversely related to the rate at which the required input mix changes.

The <u>impact of variation on V and W</u>. That T* also depends on the absolute levels of V and W can be demonstrated by differentiating (2.13) partially with respect to V and W yielding:

$$\partial j_{T*}/\partial V = EW(V-W)^{-2}, \tag{2.23}$$

and

$$\partial j_{T*}/\partial W = -EV(V-W)^{-2} \tag{2.24}$$

The right-hand side of (2.23) is positive. Since V is less than zero, an increase in country 1's skill advantage means that $\partial V<0$. These together imply that j_{T*} declines with an increase in skill advantage, which for any $h_c(t)$ implies that T* increases -- in terms of Figure 2.4, $\overline{JJ}$ obtains at a smaller value of j_T.

Consider next the impact on T* when country 1's labor disadvantage vis à vis the potential imitation country is larger for a given skill advantage. The right-hand side of (2.24) is positive, and since W is positive, a larger country 1 labor disadvantage means that $\partial W>0$. Therefore j_{T*} increases with a larger labor disadvantage, implying that given any $h_c(t)$, T* declines.

Evaluated within a static framework, (the values taken by V and W constant over time), these results indicate that for given values of V and W over time, the duration of country 1's cost advantage in any commodity will vary directly with the value taken by V and inversely with that taken by W.

Alternatively, these results can be interpreted within a context in which country 1's relative labor disadvantage is not constant over time. Consider the given pair of countries, but remove the assumption that V is constant over time. Specifically, posit that country 2 is "catching up" to country 1: V is declining (becoming a smaller negative value) over time, while W is remaining constant. In terms of (2.12), country 1's "relative" labor disadvantage is increasing over time. The impact here will be threefold: (i) the required relative initial skill elasticity will increase for new products being innovated at later points in time; (ii) the duration of the country 1's cost advantage in a given commodity continues to shrink even after having been innovated, viz., given (2.13), and assuming W constant, $dJ^*/dV = EW(V-W)^{-2}\,\frac{dV}{dt}$, which is positive since

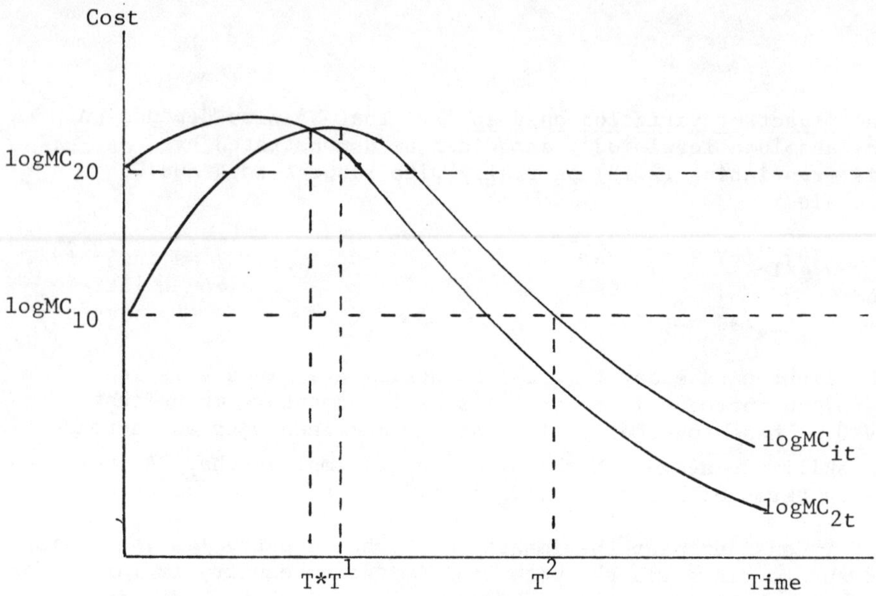

Figure 2.2
The Efficiency Criterion and the Switch in Cost Advantage

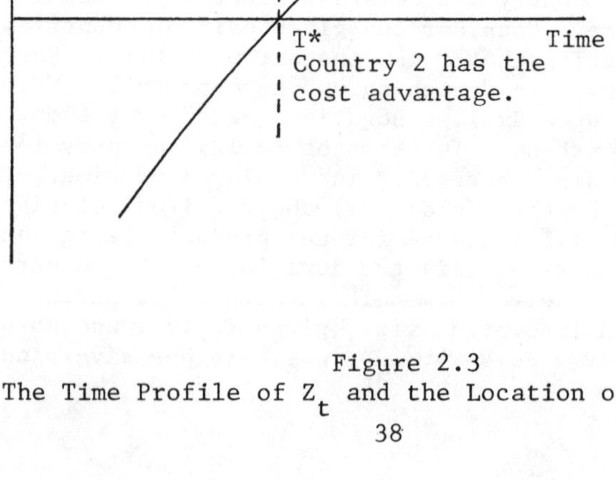

Figure 2.3
The Time Profile of Z_t and the Location of Cost Advantage

dV/dt is positive; which is to say that $\overline{JJ}$ in Figure 2.4 is positively sloped ($\overline{JJ}'$), and in the case of $h_1(t)$, the duration of country 1's cost advantage shrinks from OT_1^* to $OT_1^{*\prime}$.

Simultaneous variation in V and W. By differentiating (2.13) totally, the impact on T* of simultaneous variations in V and W (increases in both the relative skill advantage and relative labor disadvantage of country 1 in a cross-sectional context) can be examined:

$$dJ_{T*} = EW(V-W)^{-2}dV - EV(V-W)^{-2}dW = E(V-W)^{-2}(WdV-VdW) \quad (2.25)$$

Since $E(V-W)^{-2}$ is positive, the sign of dj_{T*} is determined solely by $WdV - VdW(WdV<0;\ -VdW>0)$.

Theorem: $dT^* \gtreqless 0 \Longleftrightarrow \dfrac{dW/W}{dV/V} = N_{WV} \lesseqgtr -1.$

Proof:[18]

This theorem follows directly from (2.25):

$dj_{T*} = 0 \Longleftrightarrow WdV = -VdW$, or $-1 = dW/W/dV/V$;

$dj_{T*} \gtreqless 0 \Longleftrightarrow WdV - \gtreqless VdW$ or $-1 \gtreqless dW/W/dV/V$;

since $dV<0$.

Since JT* and T* vary inversely (a smaller value of j_{T*} implies a larger value for T* for any $h_c(t)$ in Figure 2.2), the theorem has been proved.

6. On the "Pecking Order"

The import of the Theorem must not go unnoticed. It provides the necessary and sufficient condition for generating the "pecking order" of countries which is spoken of in the product cycle literature.[19] Suspend the assumption of a two-country model, and admit, besides country 1, a set of M countries. They will be ranked according to country 1's skill advantage vis à vis each m M. Let:

$V_m = \log s_1 - \log s_m;\ |V_m| < |V_{m+1}|$; all $m \varepsilon M$;

$W_m = \log w_1 - \log w_m;\ W_m < W_{m+1}$; all $m \varepsilon M$.

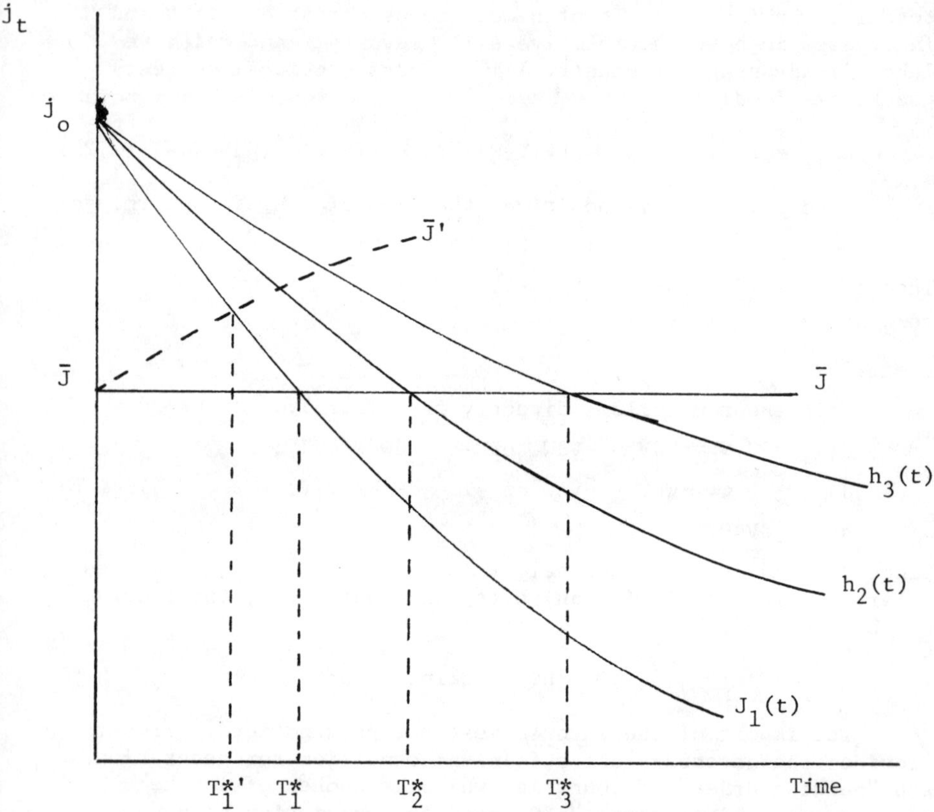

Figure 2.4

The Response of T* to Variations in
the Time Profile of the Skill-output Elasticity

The "pecking order" suggests that when considering a set of countries like Mover which country 1's relative skill advantage progressively increases, (relative labor disadvantage progressively increases), the length of time during which she will have a cost advantage vis à vis each country in any new commodity also increases: $T^*_m < T^*_{m+1}$, $\forall m \epsilon M$. Equation (2.23) taken alone suggests that this is correct: as V_m increases (in the negative direction), T^*_m also increases. But the impact of variations in W_m over the set must also be considered, and this is where the Theorem comes into play. The necessary and sufficient condition such that $T^*_m < T^*_{m+1}$ all $m \epsilon M$ obtains is that $N_{W_m V_m} > -1$ obtain for all m in M. That is, each percent increase in relative returns to skill from country m to country m+1 must be offset by less than a one percent decrease in relative returns to labor.

7. Conclusion

This chapter formalizes important aspects of Vernon's contribution to the product cycle literature. We construct a simple two-country, three-factor, multi-commodity model of dynamic comparative advantage given the innovation of one new homogeneous commodity. From this model are deduced the necessary and sufficient conditions which must be satisfied if the following assertions of the product cycle literature are to obtain:

i) initial production of the new commodity will occur in the technologically most developed country;
ii) there exists a unique point in time at which the production cost advantage shifts from country 1 to country 2;
iii) when considering a set of potential imitator countries, ranked by relative technological development, a pecking order obtains such that the length of time over which country 1 has a pair-wise cost advantage is progressively longer.

FOOTNOTES

[1]I wish to acknowledge my debt to Dr. Tom Sparrow for his invaluable assistance and guidance in the development of the analysis contained in this chapter. Any errors, of course remain the responsibility of the author.

[2]Vernon, Raymond, "International Investment and International Trade in the Product Cycle," Quarterly Journal of Economics, May, 1966.

[3]Ibid.

[4]R.W. Jones, "The role of Technology in the Theory of International Trade," mimeo, p. 15.

[5]Ibid.

[6]That a factor such as skill plays a decision role in the determination of comparative advantage is evidenced in the results appearing in: D.B. Keesing, "Labor Skills and Comparative Advantage," American Economic Review, Papers and Proceedings, May, 1966; __________, op. cit.; and Vernon & Gruber in Vernon, ed., op. cit.

[7]Setting the A's equal is done so as to abstract from the total factor productivity problem in what follows.

[8]The traditional statement of relative factor prices in situations of incomplete international factor price equalization is: $s_1/w_1<s_2/w_2$. That this is equivalent to the statement used in the derivation of (2.3) and (2.4) can be shown as follows:

$s_1r_1<s_2/r_2$ and $w_1/r_1>w_2/r_2$ together imply that

$s_1/s_2<r_1/r_2<w_1/w_2$; or $s_1/w_1<s_2/w_2$.

and when $r_1 = r_2$: $s_1/s_2<1<w_1/w_2$.

[9]Q_{Li}, Q_{Si}, and Q_{Ki} are the marginal productivities of L, S, and K respectively, and P_i is the market price of the commodity.

[10]This is a simplification which does not affect subsequent analysis, since capital is assumed internationally mobile and available to any country at an identical supply price. As a

result, the behavior of the capital-output elasticity has no effect on relative marginal costs. It is therefore assumed constant so as to be able to work with constraints which are constant over time.

[11] $b_o = E - j_o$. Substituting into (2.11a) gives
$j_o > -(E-j_o)WV^{-1} = EWV^{-1} + j_oWV^{-1}$, or
$j_o(1-WV^{-1}) > -EWV^{-1}$. And therefore $j_o > -(EWV^{-1})/1-WV^{-1}$,
or: $j_o > -EW(V-W)^{-1}$.

[12] The scale efficiency coefficient is the variable A in (2.1). Assuming this variable constant over time is tantamount to asserting that technical progress interpreted as increasing total factor productivity over time does not obtain.

[13] For a derivation of the elasticity of an isoquant, (N_{LS}), see the appendix to this chapter. The necessity of this condition is demonstrated in the discussion of sufficient conditions below.

[14] $dlogMC_{it}/dt = -db_t/dtlogb_t/dt - dj_t - dj_t/dt$
$+db_t/dtlogw_i + djt/dtlogs_i$.

Remembering that $dj_t/dt = -db_t/dt$:

$dlogMC_{it}/dt - dj_t/dtlogb_t - dj_t/dtlogj_t - dj_t/dtlogw_i + dj_t/dtlogs_i$,

[15] From (2.5) it follows that:
$logL_t = logMC_{it} + logQ_{it} + logb_t - logw_i$; therefore:
$logb_t = logw_i = logL_t - logMC_{it} - logQ_{it}$. And:
$logS_t = logMC_{it} + logQ_{it} + logj_t - logs_i$; therefore:
$logs_i = logj_t = -logS_t + logQ_{it} + logMC_{it}$. Thus:
$logb_t = logw_i + logs_i - logj_t = logL_i - logS_i$.

[16] The term "marginal technology" requires definition. As we have treated it, MC_{it} refers to the (constant) marginal cost of producing the new commodity with the technique which becomes available -- is learned -- in the current period. Merely being available, though, is not a sufficient condition for adoption of the current-period technology. Since the firm is assumed to be a profit maximizer, it will adopt that technology which minimizes marginal cost. This is the criterion for determining whether the marginal technology is superior to that currently in use: if $MC_{it} < MC_{ip}$, where MC_{ip} is the marginal cost of

presently used technology, then presumably a profit maximizing firm will adopt the new technology.

It also should be noted here that the fact that the firm does not forget previous technologies does not prohibit specifying the production function as we have. The profit maximizing firm is presumed to use that production technique which satisfies the criteria of technical and economic efficiency. While at any moment in time there exists a variety of processes for producing a given commodity, only the most efficient at the given factor prices will be adopted.

In situations where the marginal technology is always superior, firms will adopt the marginal technology, and the parameters of the production function for that commodity for that period will assume values appropriate to the adopted technique. But the fact that the firm does not forget previous techniques does not disallow the specification of a production function of the form (2.1) for the current period.

[17]The scale efficiency coefficient is assumed to be a monotone increasing function of time. Further, it is assumed that sufficient condition (2.19) is satisfied, thereby guaranteeing that marginal cost is monotone decreasing function of time.

[18]Recall that $N_{WV}<-1$ means that $0>N_{WV}>-1$.

[19]See for example: Vernon, _op. cit._, May, 1966; Hirsch, S., _The Location of Industry and International Competitiveness_, Oxford, 1967; and Hufbauer, G.C., _Synthetic Materials and the Theory of International Trade_, Harvard, 1966.

Chapter Three

The Equilibrium Time Path of Aggregate Demand for the ith New Commodity

1. Introduction

This chapter incorporates the concept of consumer learning into a model of growth in aggregate demand for new commodities. Kelvin Lancaster's "new approach" to consumer theory plays a crucial role throughout.

The model of consumer learning is presented in Section 2, while Section 3 offers a formulation of a microeconomic learning function. These are combined in Section 4 into a specification of the jth consumer's demand function for the ith new commodity. Section 5 addresses itself to the problem of aggregating the J individual demand functions into an aggregate demand function. The chapter concludes in Section 6 with the derivation of equilibrium time paths of the quantities demanded in the aggregate of the ith new commodity.

2. A Model of Consumer Behavior

If we were to base our analysis on traditional demand theory, we would be able to make little progress because traditional demand theory is incapable of enlightening us when it is confronted with a new commodity. Quandt observes, "one of the highest goals of demand estimation is to be able to predict demand for future, as yet nonexistent commodities. Traditional demand analysis is totally incapable of meeting this objective."[1] We must find then another tool of analysis, one capable of predicting the growth in demand for a new commodity. Kelving Lancaster has developed such a tool.

In his "new approach" to consumer theory, Lancaster suggests that the objective of consumers is to consume a collection of characteristics.[2] The consumer is assumed to possess a complete weak preference ordering in characteristics, resulting only indirectly in a ranking of goods bundles by virtue of the collection of characteristics they possess.

Lancaster views "goods as inputs into a process in which these characteristics are the outputs."[3] It follows from this that the typical consumption activity can be thought of as using up a unit of a commodity and thereby deriving a bundle of joint characteristics from it. This single-good-input - joint-satisfaction output is Lancaster's principle innovation.

In the "new approach" the individual's constrained utility maximization problem is written as a non-linear program. Let $X = (x_j)$ $j=1, \ldots n$ be the vector of elements denoting quantities of characteristics; $P = (p_j)$ be the vector of prices of (x_j); and $u+D(Z)$ be the ordinal preference ordering of bundles of characteristics defined on characteristics-space, (C-space). Define the "consumption technology" as the transformation between goods-space (G-space) and C-space: the system $Z=kX$ where $K=(k_{aj})$ is the matrix of objectively determined coefficients. Then the consumer's choice problem can be written:

$$\text{Maximize: } U = D(Z) \tag{3.1}$$

$$\text{Subject to: } PX \leqq c \tag{3.2}$$

$$\text{with: } Z = kX \tag{3.3}$$

To transform the budget constraint (3.2) into C-space we can write:

$$P(K^{-1}Z) \leqq c.$$

The two-goods-two-characteristics case is represented graphically in Figure 3.1. Commodities G and H embody characteristics I and II in (constant) proportions denoted by vectors Og and Oh respectively. The length of each vector is determined by the price of the commodity, (P_H, P_G), and the income of the individual (c). If total income is spent on G or H, consumption at g or h, respectively, will be realized. Given the price and income situation, the maximum amount of II that the individual can realize for any amount of I lies on gh, the efficiency frontier for characteristics, (EF).[4]

Consider a third good, W, embodying I and II in the proportion $\bar{w}$. If the price of $W_j (P_W)$, is such that $w^* \geqq \bar{w}$, the good will be sold, generating the convex EF, gw*h, and the optimal characteristics bundle will obtain along either gw* or w*h --W being substituted in consumption for one of G and J as determined by $U = D(Z)$. If the price P_W is such that only $w' < \bar{w}$ can be attained, W will not be sold, since gw'h is everywhere within gh.

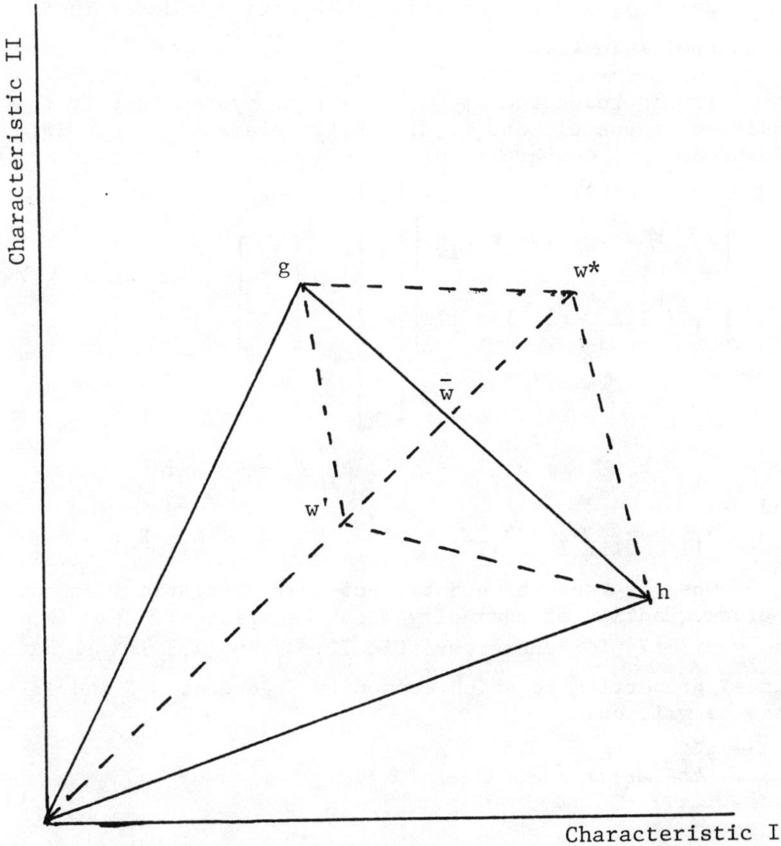

Figure 3.1

Consumer Choice in Characteristics Space

The interpretation to be given to the "corners" of the EF can be appreciated by considering the consumption technology more carefully. The latter is defined as the transformation between G-space and C-space, the system Z = KX, where:

$X = (X_j)$, j=1,. . . n; a column vector of goods;

Z=(za), a=I,II; a column vector of characteristics;

$K=(k_{aj})$, a 2 by n matrix of objectively determined coefficients.

The introduction of W alters this system only by the addition of one element to X, and two elements to K. In expanded form, we have:[5]

$$\begin{bmatrix} k_{II} \; K12 \; . \; . \; . \; k_{I}n \; K_{lw} \\ k_{III}k_{112} . \; . \; . \; k_{IIn}k_{11w} \end{bmatrix} \begin{bmatrix} X_1 \\ X_2 \\ \cdot \\ \cdot \\ \cdot \\ X_n \\ X_w \end{bmatrix} = \begin{bmatrix} 3_1 \\ 3_{II} \end{bmatrix} \tag{3.4}$$

or

$$3_I = k_{II}X_I + k_{12}X_2 +. \; . \; . \; k_{In}X_n + K_{Iw}X_wi$$

and (3.5)

$$3_{II} = k_{III}X_I + k_{II}X_2 + \; . \; . \; . \; k_{IIn}X_n + k_{IIw}X_w.$$

Observe that the quantity of characteristic I embodied in a given quantity of commodity W can be written: $3^w_1 = (k_{Iw})(X_w)$; and similarly for characteristic II: $3^w_{11} = (k_{IIw})(X_w)$. And the (true) proportion in which commodity W generates I and II can then be written:

$$\frac{3^w_{II}}{3^w_I} = \frac{k_{IIw}}{k_{Iw}} . \tag{3.6}$$

It follows that the location of a vector in C-space is determined solely by the ratio of the objectively determined coefficients. Its length is determined solely by the price of the commodity and the consumer's income; that is, the total amount of each characteristic which the consumer could realize by specializing in the purchase of this commodity. Therefore, the "corners" of the EF represent positions of complete specialization in consumption.

Given this tool of analysis, we proceed to a consideration[6] of the problem of the introduction of a new good into country 2. In addition to the above assumptions, we add the following: Prior to the introduction of the new good W, there existed three goods available in country 2: G, I*, and H; the prices of these goods remain constant. Income in 2 is constant. Each consumer in 2 possesses an ordering U=D(Z) as described above -- this ordering need not be the same for all consumers. There exist only two characteristics: I and II.

3. A New Approach to Consumer Learning

Upon introduction, the aggregate demand for a new commodity can not be expected to immediately take on its equilibrium level; consumers must be apprised of the characteristics of the new product so as to ascertain for which commodities it is a substitute, and for which one(s) it is a complement. There are three lags associated with this learning process. A lag exists before the consumers are aware of the new commodity's existence. There is a second lag while consumers become familiar with the properties and the potential uses of the new commodity. Lastly, a third lag obtains before it is fitted into the consumers' regular consumption pattern.

Interpreted within the framework of Lancaster's "new approach", learning involves discovering the true value of the elements in the column vector $(k_{a(n+1)})$ in the matrix K', (which is now 2 by (n+1)), of objectively determined coefficients in the consumption technology, where $(k_{a(n+1)})$ is the vector of coefficients associated with the n+1 good -- the ith new commodity.

Learning within this framework can be interpreted in various models. One model would suggest that the consumer will eventually know the true value of each element of $(k_{a(n+1)})$ with certainty. At the outset the consumer is presented with imperfect information, and learning involves fuller and more perfect information culminating in perfect knowledge of the value of each element in $(b_{j(n+1)})$. Thus for the ith new commodity, W, there will obtain a relation between C_{wt} and time, (where C_{wt} is the ratio of (k_{IIwt}/d_{Iwt}) at time T), such that a converging sequence will obtain. Let each of k_{IIwt} and k_{Iwt} be a sequence which converges (in the limit as T→T*) on its true value, $\hat{k}_{IIw}$ and $\hat{k}_{Iw}$ respectively. It follows that the ratio C_{wt} will will converge on its true value as T→T*.

A second model of consumer learning would take as its principle hypothesis that the consumer, given a finite time horizon, would never know the true values of these coefficients with certainty. Ignorance would be this approach be represented by letting each element in the vector, $(k_{a(n+1)})$ be a random variable, distributed with mean and variance. Learning would reduce these variances, thereby reducing the uncertainty with which each element is known. This learning would proceed via two routes: (i) over the time interval t_r to t_{r+1}, (one time period), the consumer would realize greater command of the stock of information at his disposal at time t_r, and (ii) a flow of additional information would obtain. These two forces acting in concert would lead to a reduction in the uncertainty with which the consumer knows the true value of each element in $(k_{a(n+1)}0$.

A Model of Consumer Learning in a World of Uncertainty

Assumptions:

1) Ignorance in a world of uncertainty manifests itself in the form of a density function of characteristic ratios. That is for the ith new commodity, W:

   $C_{wt} = \dfrac{K_{IIwt}}{K_{Iwt}}$ is a random variable distributed with mean

   $\hat{C}_w$ and variance s^2_{wt}.

2) $\hat{C}_w$ is the true value of the characteristic ratio and is invariant throughout the learning period.

3) The learning sequence suggested by lags of types (i) to (iii) above manifests itself in the form of a functional relationship between s^2_{wt} and time; specifically $s^2_{wt} = Q(t)$ such that $ds^2_{wt}/<dt\ 0$, and the $\lim_{t \to t_n} s^2_{wt} = \omega^7$, where ω is a positive value within a small neighborhood about zero. The "learning period" is defined as the interval $t_0 - t_n$ over which $s^2_{wt} \geqq \omega$.

4) Throughout the learning period the consumer consistently underestimates the true values of K_{IIw} and K_{Iw}. This follows from the lack of complete information during the learning period.

5) There obtains a continuous flow of w-specific information to the consumer during the learning period. This flow has a positive first derivative and a negative second derivative with respect to time. Therefore, the first and second time derivatives of the stock of w-specific information are positive and negative respectively with the former approaching zero as $t \to t_n$.

It is recognized that by restricting misestimation to underestimation, we are ruling out the case where initial over-estimation of K_{IW} or I_{IIW} occurs, and therefore are ruling out the situation where substitution effects due to learning (shortening of characteristic ratios) would work against those due to price effects.

The impact on our model of removing this restriction would appear in (3.11). The formulation appearing there would require a negative feedback term for the subset of consumers in the set j = 1, . . . J who overestimate some characteristic. If the object of our analysis was solely the specification of the micro-demand function, inclusion of such a term would be essential. But since we proceed to aggregate over the set of J consumers, we choose to beg the issue of negative feedback resulting from overestimation of K_{IW} or K_{IIW}, since we feel that aggregate overestimation would be the exception rather than the rule.

The behavior in C-space generated by this model is depicted graphically in Figure 3.2. Commodity W is introduced at time t_o. The set of feasible characteristic ratio vectors at time t_o lies within the bounds of the limiting vectors $B_O = \hat{C}_w + \#s_{wo}$, and $A_O = \hat{C}_w - \#s_{wo}$.[8] By the time $t_1 = t_o + \Delta t$, the consumer has acquired greater experience with the ith new commodity, the impact of which is to reduce the feasible characteristic ratio set to that bounded by the limiting vectors $B_1 = \hat{C}_w + \#s_{w1}$, and $A_1 = \hat{C}_w - \#s_{w1}$.

The increasing stock of w-specific information over time enables the consumer to better identify the good(s) I* for which the ith new good is a substitute, and he can be expected to increasingly include the latter in his consumption pattern at the expense of I*. This substitution is abetted from the supply side by the behavior of the market price for the ith new commodity. In Chapter Two, it was suggested that the market price of the ith new good can be expected to be a monotone decreasing function of time.[9] Such price behavior would generate substitution and income effects favorable to increased substitution of the ith new commodity for good(s) I*, since it will act to lengthen intermediate as well as true characteristic ratio vectors for any given absolute values of K_{Iwt} and K_{IIwt}. A second stimulus to substitution obtains from decreasing underestimation of the true values of k_{IIw} and K_{Iw} as learning proceeds. This follows from the observation that for a given price of w, and given relative values of the two coefficients, larger absolute values for each result in a longer vector.

In terms of Figure 3.3 the analysis proceeds as follows. Prior to the introduction of the ith new commodity, the consumer maximizes U = D(Z) subject to the consumption technology defined by the efficiency frontier gi*h; G, I*, and H being the then available commodities. At time t_o the ith new commodity becomes available. The vectors $[A_o, B_o]$ form the boundaries of the distribution of characteristic proportion vectors at t_o, while $\hat{c}_w$ is the "true" characteristic vector -- the one which obtains with a probability approaching unity as $t \to t_n \Longrightarrow s^2_{wn} \to \omega$.

During the interval $t_o t_n$, (the duration of the learning period), the consumer will make inefficient use of the new consumption technology (the EF $g\hat{C}_w h$). His objective of efficient choice is obstructed by the variance of the probability distribution of C_{wt}. Observe that at some arbitrary point

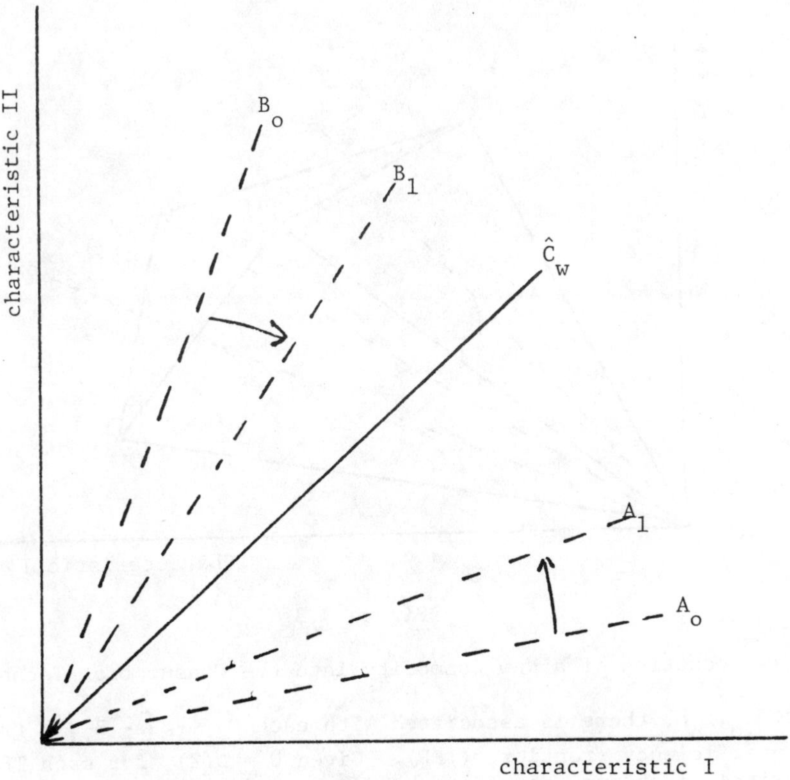

Figure 3.2

Consumer Learning and the Variance of Characteristic Vectors

$B_o = \hat{C}_w + \# s_{wo}$ $\quad\quad$ $B_1 = \hat{C}_w + \# s_{w1}$

$A_o = \hat{C}_w - \# s_{wo}$ $\quad\quad$ $A_1 = \hat{C}_w - \# s_{w1}$

In the limit as $t \to tn$ the distribution can be written as:

$$c_{wn} = C_w \underset{-}{+} \omega.$$

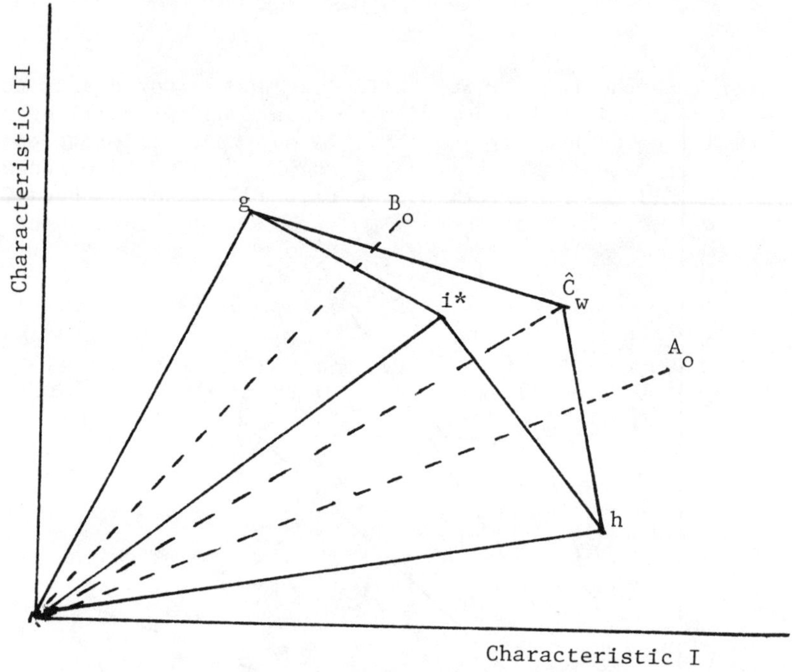

Figure 3.3

Introduction of a New Commodity into the Consumption Technology

$\bar{t}\varepsilon\ [t_o, t_n]$, there is associated with each $C_{wt}\ \text{-}\varepsilon\ [A_{\bar{t}}, B_{\bar{t}}]$ a transitory efficiency frontier, (TEF). Given U = D(Z), for each TEF there exists an optimal commodity bundle, ϕ_b, (determined by the point of tangency between a characteristic indifference curve and that TEF). Thus for each $t\varepsilon\ [t_o, t_n]$, the set $[A_t, B_t]$ generates a set of optimal bundles $[\phi_b]$. As long as $[\phi_b]$ contains more than one element, consumer choice is inefficient, since, for the given price and income situation, efficiency requires a unique optimum. By this definition of efficiency, consumer choice within the context of an uncertainty model can never be efficient. This follows from assumption (3): within a finite horizon, learning is able to reduce s^2_{wt} only to a value of ω, in the neighborhood of zero.

Thus the role of information in such a world acts to minimize the inefficiency with which a consumer utilizes a consumption technology, which is to say that efficient choice by the above definition is not possible in a world of uncertainty. It follows then that as there is a dispersion about C_w of magnitude $2\pm\sqrt{\omega}$, there must also obtain dispersions about g and h which are in the neighborhood of zero.

This result by no means vitiates the role of learning. It functions to maximize feasible choice efficiency by minimizing the number of elements in $[\phi_b]$. In a certainty model, the set $[\phi_b]$ would be reduced to a one-element set $[\hat{\phi}]$, containing only the commodity bundle which generates $\hat{C}_w$. The effect of uncertainty then is to destroy the uniqueness of this solution, while the role of information received is to minimize the impact of uncertainty.

4. The Microeconomic Learning Function

While traditional analysis would not allow us to do so, the "new approach" enables us to introduce the concept of a "learning function" in consumption. Learning, involves the sequentially improved ability of the consumer to specify $\hat{K}_{Iw}$ and $\hat{K}_{IIw}$; this ability relating positively to the cumulative consumption of the ith commodity and the cumulative stock of information.

In what follows, a consumer learning function is developed, the dependent variable of which is termed the "proportion of total information,"I(t). This term is to be a proxy for the sequentially improved ability of the consumer to specify $\hat{K}_{Iw}$ $\hat{K}_{IIw}$, which results from w-specific information received and assimulated. This learning function has the asymptotic property that in the limit, the dependent variable I(t) assumes a value of unity reflecting complete knowledge. This is exactly parallel to the asymptotic property of s^2_{wT}: $\lim_{t\to\infty} s^2_{wT} = 0$. In similar fashion for all finite t,$<I(t)<1$ obtains, as in the case with s^2_{wT}, reflecting uncertainty.

In either a world of certainty or uncertainty, we will assume that learning proceeds in a manner akin to that observed in biological growth: beginning at a very low value, accelerating, and then finally tapering off as it asymptotically approaches the upper bound of maximum size -- or in the present context, the total information level.

In describing biological growth phenomena, researchers have been quite successful with the logistic cumulative density function:

$$S(t) = \left[1+e^{-(1+gt)}\right]^{-1}; \tag{3.7}$$

where S(t) is the percent of maximal size attained at time t, and is a function of a single parameter g. The correct interpretation to be given to the parameter g is indicated solving for t in the logorithmic transformation of (3.7):

$$t = (-g)^{-1}\log\left[1-S(t)/S(t)\right] -g^{-1} \tag{3.8}$$

From (2) it follows that

$$\hat{T}=t_2 t_1 = g^{-1}\log\left[(1-S(1))S(2)\left[S(1)(1-S(2))\right]^{-1}\right]=g^{-1}\log\hat{S} \tag{3.9}$$

where S(1) and S(2) correspond to the percent of maximal size achieved at t_1 and t_2 respectively. Letting S(1) = 5% and S(2) = 95%, T can be interpreted as the growth period, the time necessary to go from 5% to 95% of full size.[10] Solving (3.9) for g and using 5% and 95% as benchmarks gives:

$$g = 5.89/\hat{T} \tag{3.10}$$

Therefore the parameter g varies inversely with the duration of the growth period, which is to say that values taken by g govern the rate of growth in S(t).

Unfortunately for the purpose of studying consumer learning, formulation (3.7) has the undesirable property that the lower bound (0%) obtains only in the limit as t approaches minus infinity, and equals .5 at t=0. We require a formulation in which I(t) is defined over the interval $[0,1] \ni I(0) = 0$. If some initial stock of information is allowed, then $I(0)=\hat{\pi}$, where $\hat{\pi}$ is a positive value in a small neighborhood about zero. I(t) is the proportion of total information available to the consumer at time t. To be admissable, such a formulation must also be a function in a parameter reflective of the rate of learning. Formulation (3.11) satisfies these requirements. In (3.11), π is interpreted as the parameter reflecting the initial stock of information.

$$I(t) = e^{-d/t} + \pi \tag{3.11}$$

The parameter d is qualitatively the same as g in (3.7). Solving the logarithmic transformation of (3.11) for t, and defining the learning period in the same manner as was the growth period above, gives:

$$\hat{T} = -d\left[(\log I(1))^{-1} + (\log I(2))^{-1}\right]. \tag{3.12}$$

Taking 5% and 95% as the benchmarks, and solving (3.12) for d, gives:

$$d = \hat{T}(3.4)^{-1}. \tag{3.13}$$

Therefore d varies directly with the length of the learning period. Since I(t) varies inversely with d, the rate of growth in I(t) varies inversely with the length of the learning period.[11] Thus d is a single parameter which determines the rate of growth in learning.

5. The Individual Demand Function

The objective of this section is to specify the jth consumer's demand function for the ith new commodity, W, first assuming the commodity is homogeneous, and then that it is non-homogeneous.

Given the assumption of homogeneity, the equation of the jth consumer's demand curve for W can be written:

$$q_{wj} = f_j(P_w, y_j, \bar{P}_k, t);$$

where: P_w is the own-price, y_j is the jth consumer's income, t is time, and $\bar{P}_k$ is the price of all other goods (assumed constant).

Three effects can be considered: a learning effect, an income effect, and a price or substitution effect. In order to analyze each of these effects, the following specification will be utilized:

$$q_{wy} = f_j(P_w, y_j, \bar{P}_k) I_j(t); \tag{3.14}$$

$$f_j(P_w, y_j, \bar{P}_k) = Q_j(y_j)^{b_j}(P)^{-a_j}, \tag{3.15}$$

$$I_j(t) = e^{-dj/t+\pi_j}; \quad (j=1. . . J). \tag{3.11'}$$

Equation (3.14) states that the learning effect combines with the income and substitution effect multiplicatively. The introduction of learning in this manner allows it to affect both the position and slope (at a given price-quantity pair) of the demand curve, rather than solely one or the other, as would be the case if it entered in an additive fashion. Equation (3.15) is a constant elasticity demand function, where $\bar{Q}_j$ is determined by the vector $\bar{P}_k$, and b_j and a_j are the income and price elasticities of demand respectively. Equation (3.11) is the jth consumer's learning function for W, in which d_j indicates his rate of learning and π_j his initial stock of w-specific information.

If we were to replace our homogeneity assumption by a non-homogeneity assumption, equations (3.14) and (3.15) would be amended:

$$q_{wj}^{2} = f_j(P_{w1}, P_{w2}, y_j, \bar{P}_k), \tag{3.14'}$$

and

$$f_j(P_{w1}, P_{w2}, y_j, \bar{P}_k) = \bar{Q}_j(y_j)^{b_j}(P_2)^{-a_j}(P_1/P_2)^{c_j}. \tag{3.15'}$$

Non-homogeneity allows separate demand functions for each variety of the ith new commodity to be written. Thus (3.14') is the jth demand curve for the 2-variety of the ith new good. Equation (3.15') indicates that the influence of the A-variety of W enters via a relative price effect, in which c_j represents the jth consumer's cross price elasticity of demand for the 2-variety of the ith new commodity.

Regardless of which homogeneity assumption is employed, the basic concept underlying the model is that of the "equilibrium" or "full knowledge" demand curve. Specify the values of y_j, $\bar{P}_k$, and P_b (for the non-homogeneous case), then the "equilibrium" demand schedule is the one which obtains when I(t)=1. Ceteris paribus, the position of the equilibrium demand schedule in price-quantity space is invariant. In less than perfect knowledge situations, sub-equilibrium schedules will obtain. As learning progresses, the sub-equilibrium demand schedules will asymptotically approach the equilibrium schedule for the jth consumer, the sequential position of the latter being determined by time profile of the jth income, the vector $\bar{P}_k$ (k=1,. . . z_j), and the prices of other varieties of the ith new commodity in the non-homogeneous case.

6. Aggregation

The behavior of this system is considered in greater detail below, but before doing so, attention must be given to the problem of aggregating the J individual demand functions into an aggregate demand function for the ith new commodity, W. From (3.14), (3.15) and (3.11'), and from (3.14'), (3.15'), and (3.11'), we can write equations (3.16) and (3.16') for the homogeneous and non-homogeneous cases respectively:

$$q_{wj} = \bar{Q}_j \left[y_j\right]^{b_j} \left[P_w\right]^{-a_j} \left[e\right]^{-d_j/t+\pi j} \tag{3.16}$$

$$q^2_{wj} = \bar{Q}_j \left[y_j\right]^{b_j} \left[P_2\right]^{-a_j} \left[P_1/P_2\right]^{c_j} \left[e\right]^{-dj/t+\pi j} \tag{3.16'}$$

In order to abstract from the problem of a functional relationship between income and income elasticity of demand, it is assumed that b_j can be replaced in (3.16) and (3.16') by b, the aggregate income elasticity of demand for the ith new commodity.

The j price elasticities of demand present a more serious problem. These are likely to be distributed throughout the economy with a non-zero variance. One means of aggregation would be to take a simple average, i.e., $\sum_j dj/J = \hat{a}_s$. But this method neglects the impact of differences in income among the j consumers. Consider an economy of ten individuals, for seven of whom $a_j = 1$, (j=1. . . 7), and for three of whom, $a_j = 3$, (j=8. . . 10). In this economy, $\hat{a}_s = 1.6$.

A second method of aggregation considers the income distribution. This is achieved by taking a weighted average of the a_j:

$$\hat{a}w = \sum_{j=1}^{J} a_j w_j \quad \sum_{j=1}^{J} w_j^{-1}; \quad \sum_{j=1}^{J} w_j = 1, \tag{3.17}$$

where w_j is the proportion of national income accounted for by the jth income. For example, assume that the first seven individuals each received 7% of national income, while the remaining three each received 17%; then $\hat{a}_w = 2.02$.

By taking into account the relative impact of each of the J elements, the weighted average approach is the superior of the two. It will reflect with greater accuracy the aggregate response to a given price change.[12]

Having considered the aggregation of individual income and price elasticities, the remaining parameters to be considered are: d_j, the rate of learning; π_j, the initial stock of information; and c_j, the cross price elasticity of demand for the non-homogeneous case. Without explicitly presenting the analysis, the aggregation of each of these is obtained in the same manner as was a_w. That is, the values taken by each in the aggregate demand functions is a convex combination of individual values, in which the jth percent of total income are used as weights.

Given that the aggregation procedure has been performed, the macroeconomic counterparts for the J microeconomic functions can be written:

$$Q_T = \bar{Q}Y^{\hat{b}}P^{-\hat{a}}e^{-\hat{d}/t+\pi}, \tag{3.18}$$

and

$$Q_{2T} = \bar{Q}Y^{\hat{b}}P_2^{-\hat{a}}[P_1/P_2]^{\hat{c}}e^{-\hat{d}/t+\hat{\pi}}, \tag{3.18'}$$

for the homogeneous and non-homogeneous commodity cases respectively.

The subscript w has been dropped in equations (3.18) and (3.18'). This is purely for notational convenience. Equations (3.18) and (3.18') refer throughout the ith new commodity.

7. Aggregate Equilibrium Time Paths: Quantities Demanded

The objective of this section is to generate equilibrium time paths of quantities demanded of the ith new commodity in the aggregate. In addition, analysis of the impact of the three effects noted above is offered, these being the learning, income, and substitution effects. The model to be used will consist of equations (3.18) or (3.18'), (3.19), and (3.20'):

$$Y = Y_o e^{vt} \; ; \tag{3.19}$$

$$P = [\hat{P}+P_o e^{-mt}] \; ; \; \lim_{t\to\infty} P=\hat{P} \; ; \; P_o = P(o) - \hat{P} \; ; \tag{3.20}$$

$$P_2 = [\hat{P}_2+P_{2o}e^{-mt}] \; ; \; \lim_{t\to\infty} P_2 = \hat{P}_2 \; ; \; P_{20} = P_2(0)-\hat{P}_2. \tag{3.20'}$$

Equation (3.19) states that national income is growing exponentially at a constant rate, v. Equations (3.20) and (3.20') describe the time profile of the price of the ith new commodity under the alternative homogeneity assumptions. These equations state that the market price declines exponentially over time asymptotically approaching $\hat{P}$ and $\hat{P}_2$ respectively. The variables P_o and P_{2o} each indicate the differential between the introductory price (at t-0) and the lower bound price P and P_2 respectively.

Supply considerations enter the model through (3.20) or (3.20'). These equations set the price to which consumer demand must optimally adjust so as to determine the equilibrium quantity demanded at each point in time.[13]

<u>The Pure Learning Time Path</u>. By holding income and all prices constant, the impact of learning on demand can be isolated. In this situation, the rate of growth in quantity demanded is determined solely by the rate of growth of learning, with the latter controlling the speed with which the sub-equilibrium demand schedule migrates toward the invariant equilibrium, or full-knowledge, schedule.

Taking the logarithmic transformation of (3.18), the equilibrium time path can be written:

$$\log Q(t)_1 = \log\bar{Q} + \hat{b}(\log Y) - \hat{a}(\log P) - \hat{d}(t+\hat{\pi})^{-1} \qquad (3.21a)$$

and letting $Q^* = \hat{b}\log Y - \hat{a}\ \log P$,

$$\log Q(t)_1 = Q^* - \hat{d}(t+\hat{\pi})^{-1} \qquad (3.21b)$$

Differentiating (3.21b)[14] twice with respect to time gives:

$$\frac{d\log Q}{dt} = \hat{d}\left[t+\hat{\pi}\right]^{-2} > 0, \text{ and} \qquad (3.22)$$

$$\frac{d^2\log Q}{dt^2} = -2\hat{d}\left[t+\hat{\pi}\right]^{-3} < 0. \qquad (3.23)$$

The rate of growth in equilibrium quantities demanded is positive but declining.

The $\lim_{t\to\infty} \frac{d\log Q}{dt} = 0$, which is to say that

$$\lim_{t\to\infty} \log Q = \log Q + \hat{b}\log Y - \hat{a}\log P = Q^* \qquad (3.24)$$

This result is illustrated in Figure 3.4 and 3.5. As learning progresses, the sub-equilibrium demand schedules asymptotically approach D*D*, the equilibrium quantity demanded at each point in time asymptotically approaches Q*. Therefore the pure learning time path is bounded by a maximal value Q* which the time path asymptotically approaches in the limit as $t \to \infty$.

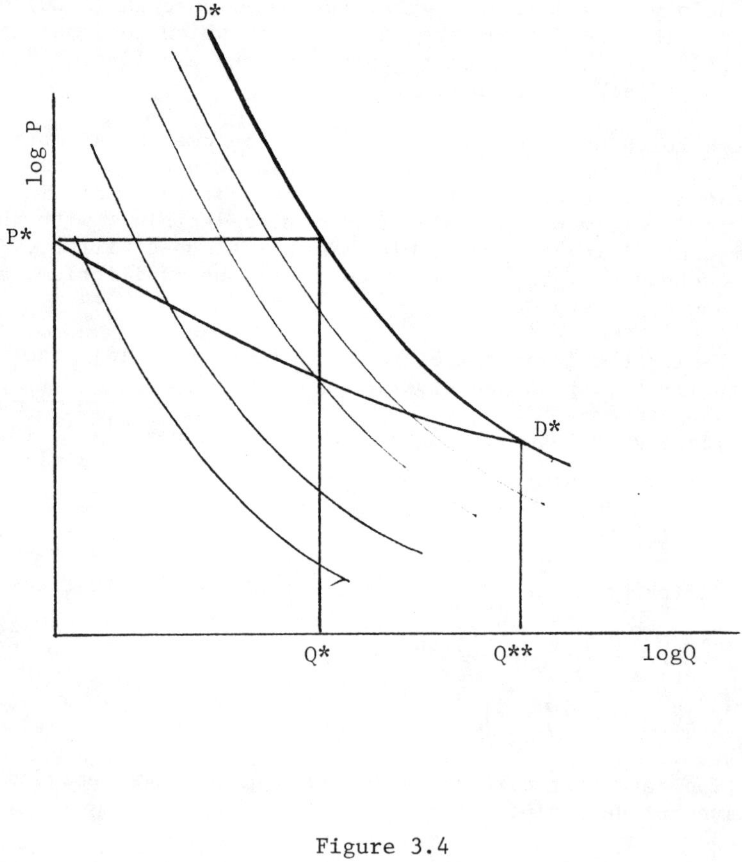

Figure 3.4

Growth in Demand for a New Commodity

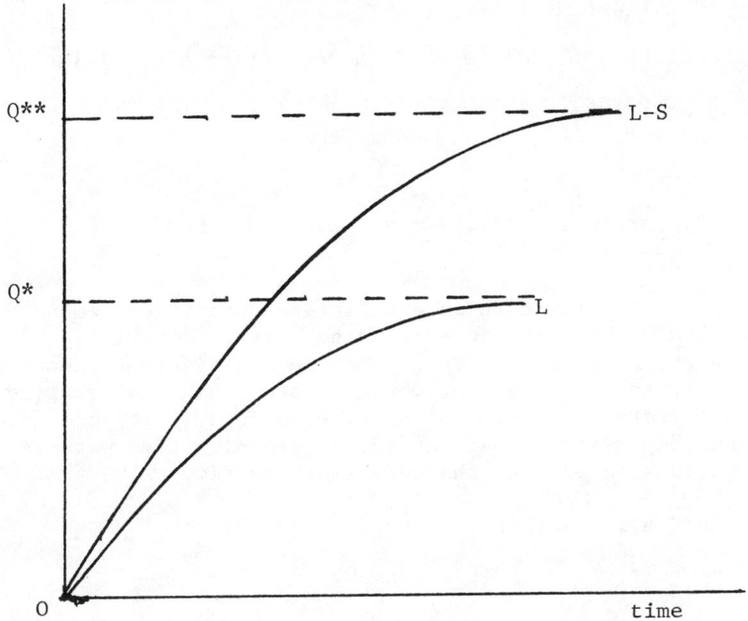

Figure 3.5

Equilibrium Time Paths

Path OL is the pure learning equilibrium time path.
Path OL-L is the learning-substitution equilibrium time path.

<u>The Learning-Income Time Path</u>. Consider a situation with constant prices and admit economic growth to the model. The latter makes the position of the equilibrium demand schedule, D*, a function of time also. Specifically, given that p=P* (In Figure 3.4) is maintained, Q* will grow with income at a rate determined by the product of the income elasticity of demand, b, and the rate of growth of the economy, v.

Substituting for Y in (3.18) the relation given in (3.19), and making a logarithmic transformation yields:

$$\log Q = \log\bar{Q} + \hat{b}\log Y_o - \hat{a}\log P + \hat{b}vt - d(t+\hat{\pi})^{-1} \qquad (3.25)$$

and the related equilibrium time path of quantities demanded is:

$$\log Q(t)_2 = Q^* + \hat{b}vt - \hat{d}(t+\hat{\pi})^{-1}. \qquad (3.26)$$

[15]

Differentiating (3.26) twice with respect to time yields:

$$\frac{d\log Q(t)_2}{dt} = \hat{b}v + \hat{d}(t+\hat{\pi})^{-2} > 0, \qquad (3.27)$$

and

$$\frac{d^2\log Q(t)_2}{dt^2} = -2\hat{d}(t+\hat{\pi})^{-3} < 0. \qquad (3.28)$$

The rate of growth in equilibrium quantities demanded is determined by both the pure learning effect, $\hat{d}(t+\hat{\pi})^{-2}$, and the pure income effect, $\hat{b}v$. These combine additively to generate a growth rate which is everywhere greater than that realized in the pure learning case. Secondly, equations (3.26) and (3.27) together demonstrate that the income-learning time path is everywhere greater than the pure learning time path, viz.,

$$\log Q(t)_2 - \log Q(t)_1 = \hat{b}vt > 0, \text{ and}$$

$$\frac{d\log Q(t)_2}{dt} - \frac{d\log Q(t)_1}{dt} = \hat{b}v > 0.$$

The most important result of admitting economic growth is that the time path is no longer bounded by a maximal value:

$$\lim_{t\to\infty} \frac{d\log Q(t)_2}{dt} = \hat{b}v, \text{ and} \qquad (3.29)$$

$$\lim_{t\to\infty} \log Q(t)_2 = t\infty. \qquad (3.30)$$

As long as the income elasticity and all other arguments of (3.18), save income, remain constant, the market for the ith new commodity will grow indefinitely at a rate which declines over time toward a lower bound of bv.

This result may appear to be at variance with empirical experience. As a commodity becomes mature of quasi-standardized, one would not expect the constancy condition to be maintained. In particular, such products tend to experience declining income elasticities of demand.[16] Secondly, in a real world context, provision must be made for the introduction of commodities which are substitutes in consumption. The latter can be expected to evidence itself in a declining value of $\log\bar{Q}$ since this variable is determined by the availability, as well as prices, of other commodities. Lastly, products do cease to exist; a result which could not obtain within the context of equation (3.26).

If the object herein were to analyze the product cycle in its entirety, these criticisms would have to be incorporated into the logic of the model. But this is not our goal; rather an attempt is being made to learn something about new products while they are relatively young. We truncate the product cycle and consider only these goods which can be classified as new and growth products[17], and with this in mind, attempt to build models reflective of the behavior of commodities of this genre.

<u>The Learning-Substitution Time Path</u>. Holding income and the prices of other goods constant, consider the impact of admitting price as well as learning effects to the model.

Under the assumption that the ith new commodity is homogeneous, the equation of the time path in logarithmic form becomes:

$$\log Q(t)_3 = \log\bar{Q} + \hat{b}\log Y - \hat{a}\log(\hat{P}+P_o e^{-mt}) - \hat{d}(t+\hat{\pi})^{-1} \quad (3.31)$$

Its rate of growth is:

$$\frac{d\log Q(t)_3}{dt} = d(t+\hat{\pi})^{-2} - \hat{a}\left[\frac{-mp_o e^{-mt}}{\hat{P}+P_o e^{-mt}}\right] > 0, \quad (3.32)$$

which is positive. The rate of growth varies according to:

$$\frac{d^2\log Q(t)_3}{dt} = 2\hat{d}(t+\hat{\pi})^{-3} - a\left[\frac{m^2\hat{P}P_o e^{-mt}}{(\hat{P}+P_o e^{-mt})^2}\right] < 0, \quad (3.33)$$

which is negative.

As in all previous cases, the growth rate of equilibrium quantities demanded is positive but declining. As in the pure learning case, there exists a maximal quantity bound:

$$\lim_{t\to\infty} \log Q(t)_3 = \log\bar{Q} + \hat{b}\log Y - \hat{a}\log P = Q^{**}.$$

Since the substitution effect is negative, the declining ith market price throughout the learning period will cause a move in the south-easterly direction along each sub-equilibrium demand schedule, as well as along the equilibrium schedule. Therefore, the learning-substitution time path lies everywhere above that for pure-learning. This is indicated in Figure 3.4.[18]

<u>The Complete Specification</u>. The impact of each of the several effects has been considered in some detail. Admitting all effects to operate simultaneously yields two equilibrium time paths; one when the ith new commodity is assumed homogeneous, and one when it is assumed non-homogeneous.

The homogeneous specification is given by (3.34) followed by the first and second-order time derivatives.

$$\log Q_H = \log\bar{Q}+\hat{b}vt+\hat{b}\log Y_o-\hat{a}\log(\hat{P}+P_o e^{-mt})-d(t+\hat{\pi})^{-1} \qquad (3.34)$$

$$d\log Q_H/dt = \hat{b}v+d(t+\hat{\pi})^{-2}-\hat{a}(-mP_o e^{-mt})(\hat{P}+P_o e^{-mt})^{-1}>0;$$

$$d^2\log Q_H/dt^2 = 2\hat{d}(t+\hat{\pi})^{-3}-a(m^2\hat{P}P_o e^{-mt})(\hat{P}+P_o e^{-mt})^{-1}<0;$$

and $\lim_{t\to\infty} d\log Q_H/dt = \hat{b}v.$

The non-homogeneous specification is given by (3.35) which is followed by its first -- and second -- order time derivatives.

$$\log Q_N = \log\bar{Q} - (\hat{d}/t+\hat{\pi}) + \hat{b}\log Y_o + \hat{b}vt - \hat{a}\log P_2 - \hat{c}\log P_2 + \hat{c}\log P_1 \qquad (3.35)$$

but $P_2=\hat{P}_2 + P_o e^{-mt}$,

and letting $g=a+c$:

$$\log Q_N = \log\bar{Q} - \hat{d}(t+\hat{\pi})^{-1} + \hat{b}vt + \hat{b}\log Y_o - g\log(\hat{P}_2+P_o e^{-mt})+\hat{c}\log P_1 \qquad (3.36)$$

$$d\log Q_N/dt = \hat{d}(t+\hat{\pi})^{-2} + \hat{b}v - g(-mP_o e^{-mt}/\hat{P}_2+P_o e^{-mt})>0;$$

$$d^2\log Q_N/dt = -2\hat{d}(t+\hat{\pi})^{-3} - g(m^2\hat{P}_2P_o e^{-mt})(\hat{P}_2+P_o e^{-mt})^{-2}<0;$$

and $\lim_{t\to\infty} d\log Q_N/dt = \hat{b}v.$

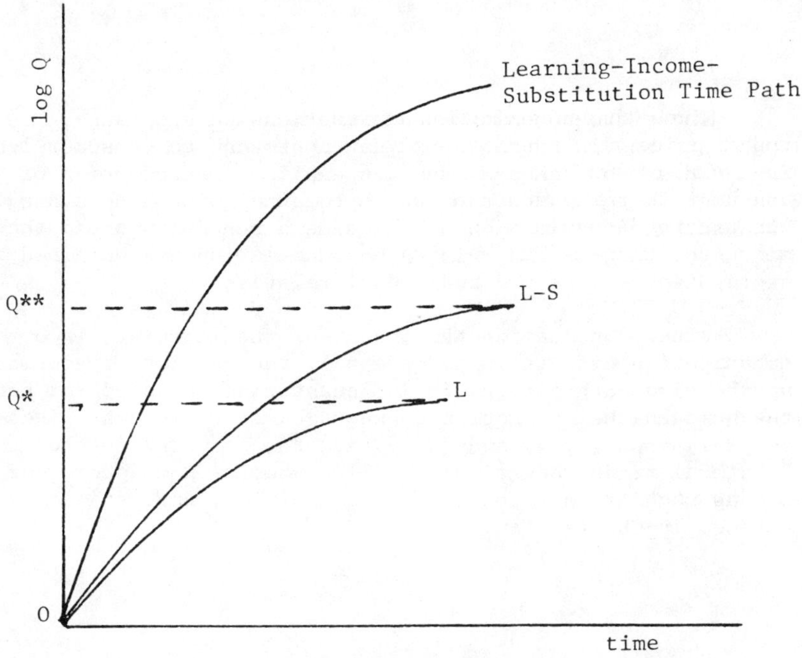

Figure 3.6

Equilibrium Time Paths of Demand for a New Commodity

The impact of the several effects individually and in combination is summarized in Figure 3.6. When only the learning effect is operational, the maximal market size for the new commodity is limited to Q*. Allowing the learning and substitution effects to work in concert increases the maximum market size to Q**.

Admitting economic growth removes the upper bound on the size of the ith market. The market can grow forever at a rate which approaches a lower bound determined by the product of the income elasticity of demand and the rate of growth of the economy, bv.

8. Conclusion

In this chapter, the concept of consumer learning was incorporated into a model of the growth in aggregate demand for the ith new commodity.

The microeconomic demand functions were obtained via an appeal to Kelvin Lancaster's "new approach" to consumer learning. This approach admitted the explicit introduction of a consumer learning function which together with a constant elasticity specification of the demand schedule formed the J micro-relations. The related macro relation was obtained by aggregating over the J individual relations.

After considering the impact of the learning, income, and substitution effects in some detail, the complete specification of the time path of equilibrium quantities demanded for both the homogeneous and non-homogeneous cases was given. These two relations (3.34) and (3.35) are the cornerstones for the bilateral trade models which will be built in Chapter Four, and as such are also basic to the empirical analysis to be pursued in Chapter Five.

FOOTNOTES

[1]R.E. Quandt, "The Construction of Travel Demand Models with Incomplete Data," Mimeo, p. 10.

[2]K. Lancaster, op. cit., May, 1966.

[3]Ibid., p. 14.

[4]Ibid., p. 17.

[5]This demonstrates the assertion that introduction of a new commodity involves no loss of information: the k_{aj} $(j \neq w)$ remain both unchanged and still useful.

[6]We are assuming a two-country model (as in Chapter one), were country I innovates the commodity, and subsequently sells it in country 2.

[7]The essence of this assumption is that perfect (or "complete") knowledge is unattainable within a finite time period in a world of uncertainty. It would not be unreasonable to suppose that the $\lim_{t\to\infty} s^2_{wt} = 0$.

[8]The parameter # is chosen so that there is a 99% probability that the vector chosen falls within the interval B_t, A_t.

[9]It should be reemphasized here that market prices could increase because of decreasing returns to scale and rising factor prices, factors which are ruled out by the model of Chapter Two.

[10]The choice of 5% and 95% as the benchmarks over which the growth period is measured are arbitrary; one could just as well have chosen 1% and 99%, but doing so of course would increase both T and $\log \hat{S}$ thereby not fundamentally altering the value taken by g. But once chosen, the benchmarks must be maintained throughout the analysis.

[11]Differentiating (3.11) with respect to d gives:

$$dI(t)/dd = -e^{-d/y+\pi} < 0.$$

[12]The weighted average and simple average approaches generate the same result in two situations: (i) $y_j = y_i \forall j \neq i$, J incomes are the same; and (ii) $a_j = a_i \; \forall j \neq i$, that is when all values of the variable being averaged are the same.

[13](3.20) and (3.20') reflect the analysis of Chapter Two which suggested that, though constant during any time period, the market price of a new commodity can be expected to be a monotone decreasing function of time.

[14]In considering the several cases, the results for non-homogeneous goods are not presented since they are exactly parallel. The complete specification of the equilibrium true path for the non-homogeneous goods case is given below.

[15]Q^* in (3.26) is equal to $\log\bar{Q} + \hat{b}\log Y_o - \hat{a}\log P$.

[16]cf., Houthakker and Magee, "Income and Price Elasticities in World Trade," _Review of Economics and Statistics_, May, 1969. They found that the income elasticity of demand for the exports of several non-industrial countries (particularly those of Latin America) to be quite los: India, .54; Brazil, .34; Columbia, .41; Argentina, .87; Chile, .99. These countries characteristically export primary commodities.

[17]cf., Seev Hirsch, _op. cit._

[18]Subtracting (3.24) from (3.31) yields:
$a \log P - \log(\hat{P} + P_o e^{-mt})$. Since $P_o = P - \hat{P}$, this can be written: $a \log P \log(Pe^{-mt} + \hat{P}(1 - e^{-mt}))$, which is positive, since $P > \hat{P}$. Thus $\log Q(t)_3 > \log Q(t)_1 \ \forall t > 0$.

FOUR

A Product Cycle Model with a Non-Homogeneous Commodity

1. Introduction

The objective of this chapter is to construct a product cycle model which assumes new products to be non-homogeneous. Section 2 addresses itself to what might be termed the "Posner-Vernon-Hufbauer" result, and indicates that it can be derived from a simple model which, while capturing the spirit of previous chapters, assumes that all new commodities are homogeneous. Section 3 examines a model of bilateral trade in the ith new commodity under the assumption that the latter is non-homogeneous. In Section 4, the results of a simulated study of the bilateral trade model are presented.

2. The Posner-Vernon-Hufbauer Result

A recurring theme of the product cycle literature is that given the genesis of the innovation, the innovating firm enjoys a period of monopoly production, but in time, production tends to be transferred to the import market, causing the innovating firm's exports to diminish. In the end, this external trade flow either disappears (Posner[1]), or changes direction (Hufbauer, Vernon), with the previous export market now becoming the market of import.[2]

One possible homogeneous-new-good product cycle model is given by equations (4.1) and (4.2). Q_{i2t} is domestic demand for the ith new commodity in country 2 (the imitating country); S_{i2t} is domestic supply of the ith new commodity in country 2, (at time t).

$$Q_{i2t} = t^a ; \tag{4.1}$$

$$S_{i2t} = t^b - c ; \tag{4.2}$$

and

$b>a>0$;

$S_{i2t} = 0 \; \forall t<T_L$, T_L being the date of first production in country 2;

c>0 is a parameter which serves as a proxy for both relative technological sophistication and product complexity.[3]

The constraints put on the functions are: that they be positively sloped; that the supply function intersect the demand function once, from below; that they be single-valued in the dependent variable; and lastly, that they be both continuous and smooth. There are no transportation costs.

Given (4.1) and (4.2), we can express the export flow from country 1 (the innovating country) to country 2 at time t as:

$$X_{122t} = \begin{cases} t^a \; \forall \, t \leqq T_L \\ t^a - t^b + c \forall t > T_L \end{cases} \quad (4.3)$$

An important concept here is that of potential foreign supply: the difference between home market demand and domestic supply of a new commodity. As long as domestic firms cannot supply the entire domestic market, national demand (in country 2) will be satisfied in part by imports from country 1, and the potential foreign supply of country 2, $\hat{X}_{i21t}$, will be negative. In terms of Figure 4.1a, the potential foreign supply of country 2 is negative for all t less than T*, and positive thereafter. In terms of our model, X_{i21t} is given by subtracting (4.1) from (4.2):

$$\hat{X}_{i21t} = t^b - t^a - c.$$

The transfer of production to country 2 then carries with it certain eventual loss of this market by the innovating firm, T* in Figures 4.1a and 4.1b. Since the ith new commodity is assumed homogeneous, once national firms are able to satisfy domestic demand, the market share of the innovating firm will go to zero.

The innovating firm's home-market share could be threatened as well. If factor endowments are identical in both countries, the Posner result obtains: external trade is zero for all t T*, and national firms in both markets will have to tailor their supply to their home markets. If factor endowments are not identical, and country 2 is the technologically

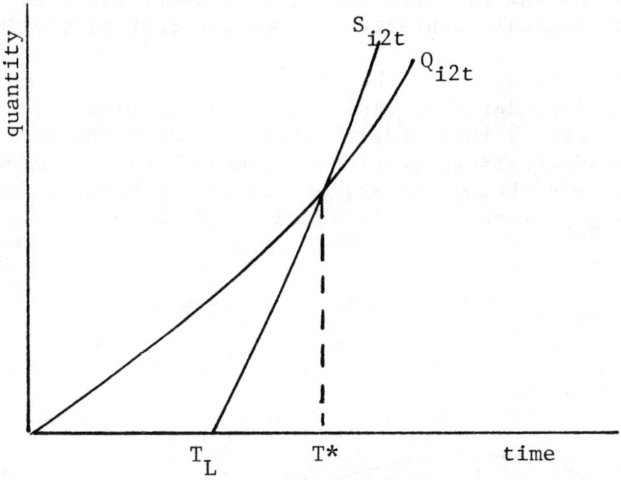

Figure 4.1a

Time Paths of Supply and Demand, the ith New Commodity

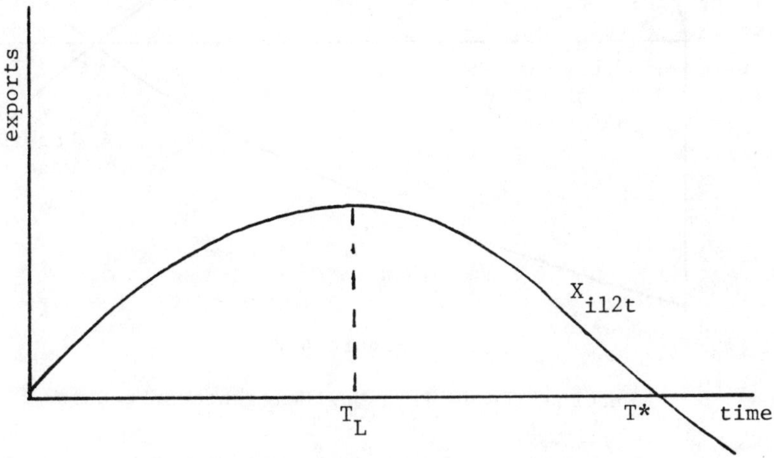

Figure 4.1b

Export Time Path of the ith New Commodity

See Appendix 5.1 for a proof that X_{i12t} is maximized at T_L.

less developed of the two, then according to the sense of the product cycle, low-wage exports will replace technological gap exports.

Figure 4.2 depicts the time profile of external trade within the context of this model. This is what might be termed the Posner-Vernon-Hufbauer result: a homogeneous-good model with external trade being mono-directional at any one time; technological gap trade being followed by low-wage trade.

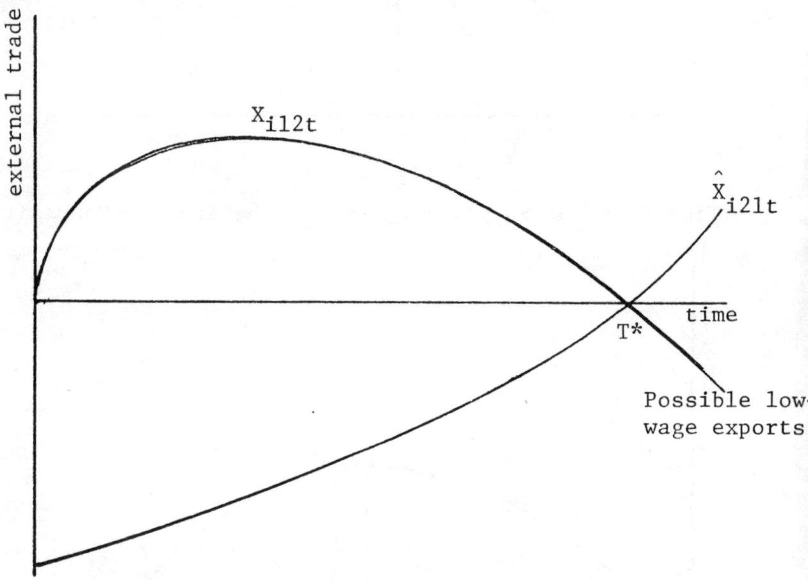

Figure 4.2

External Trade in a New Homogeneous Product

3. Non-homogeneous Commodities and the Produce Cycle

If the above correctly reflected the real world, we would not expect to observe bilateral external trade in a new commodity, but we do. This suggests that new commodities can be differentiated. As has been noted by Balassa, the majority of manufactured commodities traded by industrial countries are differentiated products.[4]

It would appear that the assumption of homogeneity of new commodities fails to capture the full essence of product cycle phenomena. External trade in non-homogeneous new commodities will be monodirectional solely during the period when only one country is producing it. As soon as both countries are each producing their own variety of a non-homogeneous new commodity, there is a potential for two-way external trade.

How must the product cycle theory be amended to account for non-homogeneity? In other words, what is the nature of the time profile of the balance of trade in the ith new commodity within the context of a two-country model, assuming that the ith new commodity is non-homogeneous?

Our central assumption in what follows is that non-homogeneity of new commodities allows us to write separate demand functions for each country's variety of the ith new commodity.

Our additional assumptions follow:

1) There exist two countries: 1 and 2.

2) Country 1 is (i) technologically superior to country 2 ($GDP/cap_1 > GDP/cap_2$)[5]; (ii) larger than country 2 ($GNP_1 > GNP_2$).

3) Each economy consists of N sectors of which the Nth is the single new non-homogeneous commodity industry; the latter being so small vis à vis the N-1 industries that it affects neither national factor nor goods prices.

4) The market price of country 1's variety of the ith new commodity is taken as the numeraire, (since it is relative, not absolute, prices that matter).

<u>Import Time Paths.</u> Using the results of Chapter Three, particularly (3.35) and (3.36), and recalling the assumption that the price of the country 1 variety of the ith new commodity is being

taken as a numeraire, we can write country 2's dynamic import demand function as:

$$M_{2t} = G(P^r_{2t}, Y_{2t}, \bar{P}_j)I(t)$$
$$= \bar{Q}_2 Y^b_{2t}(P^r_{2t})^c(e^{-(d/t+e')}) \qquad (4.4)$$

Where:

b: the elasticity of demand with respect to variations in the national income in country 2;

Y_{2t}: country 2's national income at time t;

P^r_{2t}: the relative price of the two varieties of the ith new commodity at time t;

c: country 2's elasticity of demand with respect to variations in relative prices;

d: country 2's rate of growth of learning (specific to 1's variety of the ith new commodity);

e': country 2's initial stock of information concerning the country 1 variety of the ith new commodity.

The time profile of national income in country 2 is given by (4.5). Y^o_2 is the (real) value taken by national income in country 2 at some base period, t=0, which grows exponentially at a (constant) rate, v_2.

$$Y_2t = Y^o_2 e^{v_2 t} \qquad (4.5)$$

From (3.20'), the absolute price of the country 2 variety will behave over time according to:

$$P_{2t} = \hat{P}_2 + P^o_2 e^{-mt}, \ m>0. \qquad (4.6)$$

(4.6) can be described graphically as in Figure 4.3. Since $dP_{2t}/dm>0$, $m_1<m_2<m_3$, which is to say, the larger the rate of decent, the sooner the lower limit, $\hat{P}_2$, will be realized.[6]

But we are interested here in the time profile of the relative price of country 2's variety when the market price of

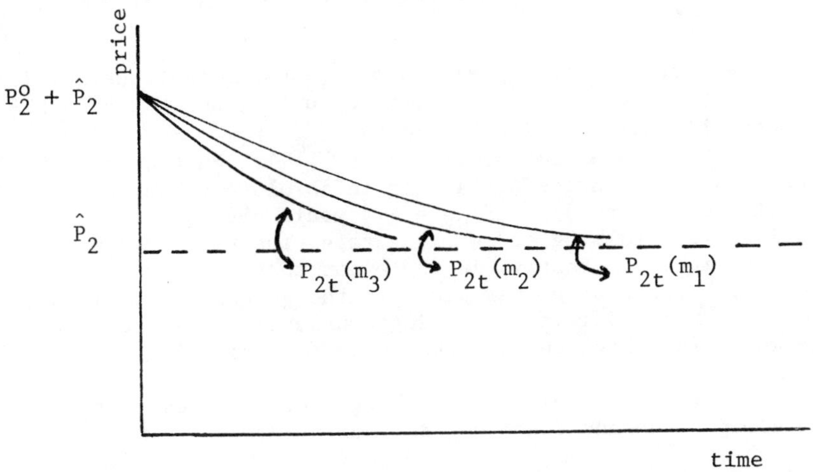

Figure 4.3

Market Price Time Profile

the country 1 variety is treated as the numeraire. This can be achieved through a modification of (4.6).

$$P_{2t}^r = \hat{P}_2^r + \hat{P}_2^{or} e^{-m't}. \tag{4.7}$$

Where:

$\hat{P}_2^r = \hat{P}_2/\hat{P}_1$, and when all necessary and sufficient conditions for a switch in comparative advantage are satisified, it follows from the analysis of Chapter Two that $\hat{P}_2^r < \hat{P}_2$, since one of the necessary conditions for a switch is that $MC_{1t} = MC_{2t}$ for some t, and that $MC_{1t} > MC_{2t}$ obtains thereafter.

$\hat{P}_2^{or}$ is the price at which the country 2 variety could be produced when the commodity is innovated by country 1 relative to the initial price of the country 1 variety.

m' is the rate of decent of P_{2t} modified for the time profile of the market price of country 1's variety.

Refer to Figure 4.4, and assume that $m = m_1$, (from Figure 4.3).

For the sake of ease of exposition only, assume that the initial market price of country 1's variety was unity, so that $P_2^o = P_2^{or}$. If country 1's market price is constant, then $m = m'$, and P_{2t}^r will follow the same time profile as P_{2t}. If, on the other hand, country 1's market price is falling in absolute value, then a path akin to (2) or (3) in Figure 4.4 would obtain. If the two prices were falling at the same rate, the time path would become horizontal. Lastly, in the event that country 1's market price was rising over time, (although unlikely, given that only new commodities are being considered), the relative price would fall more quickly than path (1), say (4) or (5).

Substituting (4.5) and (4.7) into (4.4), gives us the dynamic import demand function for country 2:

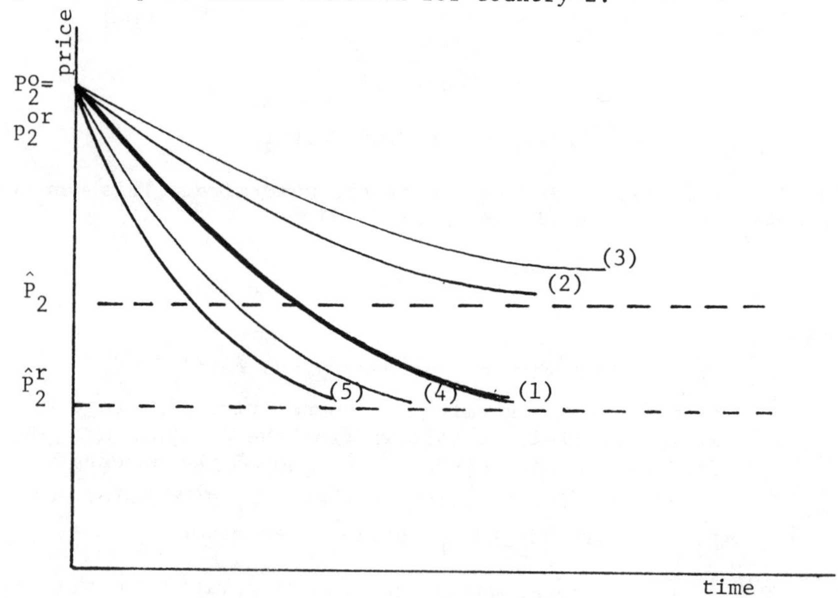

Figure 4.4

Time Profile of Relative Prices

$$M_{2t} = \bar{Q}_2(Y^o e^{v_2 t})^b(\hat{P}_2^r + P_2^{or}e^{-m't})^c(e^{-(d/t+e')}). \qquad (4.8)$$

The logarithmic transformation of (4.8) gives:

$$\log M_{2t} = \log\bar{Q}_2 + b\log Y_2^o + bv_2 t + c\log(\hat{P}_2^r + P_2^{or}e^{-mt}) - d(t+e')^{-1} \qquad (4.9)$$

Country 1's dynamic import demand function for country 2's variety can be derived in similar fashion:

$$M_{it} = H(Y_{1t}, 1/P_{2t}^r, \bar{P}_j)I_2(t) = \bar{Q}Y_{11}^f(1/P_{2t}^r)^h e^{-(o/t+z)}; \qquad (4.10)$$

$$= \bar{Q}_1(Y_1^o e^{v_1 t})^f(\hat{P}_2^r + P_2^{or}e^{-m't})^{-h}(e^{-(o/t+z)}); \qquad (4.11)$$

or

$$\log M_{it} = \log\bar{Q}_1 + f\log Y_1^o + fv_1 t - h\log(\hat{P}_2^r + P_2^{or}e^{-m't}) - o(t+z)^{-1}. \qquad (4.12)$$

Where:

- v_1 country 1's national income growth rate;
- f 1's income elasticity of demand for the country 2 variety;
- h relative price elasticity of demand in country 1;
- 0 country 1's rate of growth of learning (specific to 2's variety of the ith new commodity);
- z country 1's initial stock of information concerning the country 2 variety of the ith new commodity.

<u>Bilateral Trade Balance Time Path</u>. Subtraction of (4.12) from (4.9) gives the reduced-form equation of the time path of the bilateral balance of trade in the ith new commodity:

$$TB = \log M_{it} - \log M_{it} = N + (bv_2 - fv_1)t + (c+h)\log(P_2^r + P_2^{or}e^{-m't}) - d(t+L+e')^{-1} + o(t+z)^{-1}; \qquad (4.13)$$

where:

$$N = \log\bar{Q}_2 - \log\bar{Q}_1 + b\log Y_2^o - f\log Y_1^o + bv_2 L.$$

$$dTB/dt = bv_2 - fv_1 + (c+h)(m'P_2^{or}e^{-m't}/\hat{P}_2^r+P_2^{or}e^{-m't})+d(t+L+e')^{-2}$$
$$-o(t+z)^{-2}. \qquad (4.14)$$

Notice that in (4.13) and (4.14), the time variable in country 2's import function has been written t+L. This puts each function into its proper time sequence. At some point in time, T_o (=0), exports from country 1 to country 2 began; therefore M_{2t} is defined for all positive t. At some later point in time, T_L, country 2 exports to country 1 are initiated. Hence the time variable must be scaled so that when the reverse flow of exports begins, time in (4.8) is zero, and in (4.10) is L, the number of time periods during which external trade has been mono-directional from 1 to 2. Therefore Y_2^o and P_2^{or} are to be interpreted as the values taken when T=L, the date of initiation of the reverse flow of exports.

4. Computer Simulation

Given (4.13), we have a fully articulated model of the dynamic behavior of bilateral trade in the ith new commodity when the latter is assumed to be non-homogeneous. So that this model might be tested empirically, and also to test its sensitivity to variations in the parameters, a simulation study was completed. This study involved two steps: the first step was to solve (4.13) and (4.14) for the bilateral trade balance and its derivative for the fifty time periods, and twenty-eight iterations of the parameters. The second step was to use least-squares techniques to estimate the growth rate of the bilateral trade balance for each of the twenty-eight arrays so generated, and to test the hypothesis that (4.13) can be approximated by a log-linear specification with time as the independent variable.

For the computer simulation, the constant term, in (4.13) was deleted. Since this is a monotone transformation, it affects only the constant term, and not the slope of the regression line to be fitted by the subsequent least squares regressions. The determination of the slope is our primary concern; thus this deletion is one of convenience rather than substance.

The values taken by the parameters of (4.13) and (4.14) in each iteration are given in Table 4.1. In the first two columns appear the growth rates of national income for 2 and 1 respectively. Observe that in (4.13), for given values of the income elasticities, b and f, equal increases in both growth rates will have a neutral effect on the trade balance, which is to say that only variations in relative growth rates will influence the trade balance at any given time for given values of the income elasticities. For this reason, the growth rate of country 1 is varied so as to obtain variations in relative growth rates. The resulting interval of variation in relative growth rates is: $1 < v_1 / v_2 < 3$.

The values taken by the income elasticities for 2 and 1 respectively appear in columns 3 and 4 of Table 4.1. In order to determine within what limits to set the income elasticities, an appeal was made to the work completed by Houthakker and Magee.[7] They found that for the period 1951-1966, the United States' income elasticity of demand for imports by country varied between .75 (Netherlands) and 3.52 (Japan). Conversely, the income elasticity of demand for United States exports varied by country between .94 (South Africa) and 2.44 (Australia).[8] We have adopted these limits for the purposes of our computer simulation, the United States' import income elasticities for country 1, and her export income elasticities for country 2.

The rates of learning for 2 and 1 appear in columns 10 and 11, respectively. As with the growth rates, it is the relative values taken by the rates of learning that affect the trade balance at any given time. To this end, the rate of learning for 1 was held constant and that of 2 allowed to vary. The interval of variation in relative rates of learning is $.5<d/s<2.0$; that is 1's relative rate of learning varies from one-half of 2's to twice 2's rate.

Lastly the levels of the initial stock of information appear in columns 12 and 13 for 2 and 1 respectively. Since it is also the case here that the relative and not the absolute values which affect the trade balance, 2's initial stock of information relative to 1's varies between 30 percent of 1's to 1.5 times greater than 1's.

The computer simulation showed that the bilateral trade balance is a monotone decreasing function of time: its derivative is everywhere negative. This result concurs with the supposition of Hufbauer that low-wage trade will supplant the technological gap trade. But whereas in his model -- with its implicit assumption of new-commodity-homogeneity -- the technological gap flow disappears and is replaced by the low-wage flow, our model predicts that the innovating country will experience a continuous deterioration in its balance of trade in a new commodity once the low-wage flow has been initiated. It may turn negative, but for all values of the parameters chosen, its export flow does not disappear -- at least for fifty time periods.[9]

In order to test the hypothesis that this model can be approximated by a log linear formulation with time as the independent variable, and to test the sensitivity of the growth rates of the balance of trade to variations in the parameters, the observations generated by the computer simulation were submitted to least squares estimation. The specification used is given by (4.15):

$$\log B_{qt} = R_{q1} + R_{q2}t + u_{qt}. \tag{4.15}$$

B_{qt} is the balance of the qth iteration at time t; R_{q1} is a constant term; R_{q2} is the growth rate of the balance of trade; and u_{qt} is an error term. Regressions were run for sample sizes of 10, 12, 15, 25, and 50 time periods. Since there was little variation in the growth rates (the coefficient of the time variable) among sample sizes, and since the fit was excellent in all cases, only the results for the 50 period sample are presented.

These results appear in Table 4.2. Looking through the twenty-eight iterations, we note that the fit as measured by the R^2 is quite good. (The lowest value obtained is .975.) The regression coefficients of the time variable are all negative and strongly significant, with all coefficients being at least fifty times greater than their standard errors. This result is taken as a positive test of our hypothesis that the model can be approximated for by a log linear specification for the purposes of empirical testing of the model. The results of the latter are presented in Chapter 5.

The values taken by the growth rates vary between a low of -.025 (iteration 6) to a high of -.162 (iteration 23); which is to say that depending upon the values taken by the

Table 4.1

The Values of the Parameters Used in the Computer Simulation

#	v_2	v_1	a	f	c	g+h	m	P_2	P_o	d	s	e	z	L
1	.03	.05	1.0	1.0	.3	1.3	.02	2.0	5.0	6.0	4.0	.25	.50	10
2	.03	.09	1.0	1.0	.3	1.3	.02	1.0	5.0	6.0	4.0	.25	.50	10
3	.03	.06	1.0	1.0	.3	1.3	.02	1.0	5.0	6.0	4.0	.25	.50	10
4	.03	.06	1.0	1.5	.5	1.3	.02	1.0	5.0	6.0	4.0	.25	.50	10
5	.03	.06	1.5	1.0	.5	1.3	.02	1.0	5.0	6.0	4.0	.25	.50	10
6	.03	.03	1.0	1.0	.5	1.3	.02	1.0	5.0	6.0	4.0	.25	.50	10
7	.03	.06	1.0	1.0	.3	2.0	.05	1.0	5.0	6.0	4.0	.25	.50	10
8	.03	.03	1.0	1.2	.1	.75	.05	1.0	5.0	6.0	4.0	.25	.50	10
9	.03	.03	1.0	1.2	.1	1.3	.02	1.0	5.0	7.0	4.0	.15	.50	10
10	.03	.03	1.0	1.2	.1	1.3	.02	1.0	5.0	5.0	4.0	.30	.50	10
11	.03	.05	1.0	1.0	.3	1.3	.02	1.0	5.0	6.0	4.0	.25	.50	5.0
12	.03	.05	1.0	1.0	.3	1.3	.02	1.0	5.0	6.0	4.0	.25	.50	8.0
13	.03	.05	1.0	1.5	.3	1.3	.02	1.0	5.0	6.0	4.0	.25	.50	10
14	.03	.05	1.0	1.0	1.0	1.3	.02	1.0	5.0	6.0	4.0	.25	.50	10

Table 4.1 (Continued)

#	v_2	v_1	a	f	c	g+h	m	P_2	p_o	d	s	e	z	L
15	.03	.05	1.0	1.0	1.5	1.3	.02	1.0	5.0	6.0	4.0	.25	.50	10
16	.03	.05	1.0	1.0	1.5	2.0	.02	1.0	5.0	6.0	4.0	.25	.50	10
17	.03	.05	1.0	1.0	1.5	1.3	.02	1.0	5.0	7.0	4.0	.25	.50	10
18	.03	.05	1.0	1.0	1.5	1.3	.02	1.0	5.0	8.0	4.0	.25	.50	10
19	.03	.05	1.5	1.0	1.5	1.3	.02	1.0	5.0	6.0	4.0	.25	.50	10
20	.03	.05	.75	1.0	1.5	1.3	.02	1.0	5.0	6.0	4.0	.25	.50	10
21	.03	.05	2.0	1.0	1.5	1.3	.02	1.0	5.0	6.0	4.0	.25	.50	10
22	.03	.05	2.0	2.0	1.5	1.3	.02	1.0	5.0	6.0	4.0	.25	.50	10
23	.03	.05	1.0	3.0	1.5	1.3	.02	1.0	5.0	6.0	4.0	.25	.50	10
24	.03	.08	2.0	2.0	1.5	1.3	.02	1.0	5.0	6.0	4.0	.25	.50	10
25	.03	.05	2.0	2.0	1.5	1.3	.02	1.0	5.0	6.0	4.0	.75	.50	10
26	.03	.05	2.0	2.0	1.5	1.3	.02	1.0	5.0	6.0	4.0	.15	.50	10
27	.03	.05	2.0	2.0	1.5	1.3	.02	1.0	5.0	6.0	4.0	.25	.50	10
28	.03	.05	2.0	2.0	2.5	1.3	.08	1.0	5.0	6.0	4.0	.25	.50	10

parameters of the model, an innovating country can expect the deterioration in its trade balance with the imitating country to vary between 2.5 percent per annum and 16 percent per annum, (assuming that one time period is equal to one year).

The model appears to be quite sensitive to variations in relative income growth rates. In fact, it would appear that there is almost a one-to-one relationship between relative income growth rates and the rate of growth in the bilateral balance of trade. In iteration 2, relative income growth is 1.8 times greater than in iteration 1 -- all other parameters retain the same values. The absolute value of the rate of growth in the bilateral trade balance in iteration 2 is 1.9 times greater than in iteration 1. In iteration 3, relative income growth is 1.2 times greater than in iteration 1, and the absolute value of the resulting growth rate in the trade balance is 1.22 times larger than in iteration 1. In iteration 6, country 1's relative income growth rate is .6 of the iteration 1 level, and the resulting trade balance growth rate declines to a level of .62 of its value in iteration 1.

Variations in income elasticities also appear to have a siginificant impact on the behavior of the balance of trade. Comparing iterations 13 and 1, country 1's income elasticity of demand is .50 times greater in iteration 13. This 50 percent increase in income elasticity has resulted in a 56 percent increase in the (absolute value of the) growth rate of the trade balance. Comparing iteration 19 with iteration 21, 30 percent increase in country 2's income elasticity results in a 50 percent decline in the (absolute value of) trade balance growth rate.

The behavior of the model with respect to variations in the imitation lag bears mention. A comparison of iterations 11 and 12 with 1 indicates that the model is not terribly sensitive to variations in the duration of the imitation lag. But the results are interesting in that they indicate that a shortening of the imitation lag reduces the rate at which the trade balance deteriorates. This might at first glance appear paradoxical, but in Chapter 5, evidence is offered in support of the hypothesis that the duration of a country's imitation lag tends to vary inversely with its relative (to the innovating country) technological sophistication.[10] Bearing this in mind, it follows that the rate of deterioration of the innovating country's trade balance should vary inversely with the duration of the imitation lag, since the imitating country with a shorter imitation lag will also have less of a relative labor advantage to exploit.

The model is also not very sensitive to variation in the relative rate of learning. In iterations 17 and 18 respectively, country 1's relative rate of learning is 1.6 and 1.34 times greater than it is in iteration 15, yet the growth rate of the trade balance varies only slightly: it increases by 1 and 3 percent respectively.

The model is even less responsive to variation in the initial stock of information. In iteration 25, the initial stock of information is five times greater than in iteration 26, and yet the growth rate of the trade balance does not vary between the two iterations.

The model is responsive to variations in the price elasticities. In iteration 15, country 2's relative price elasticity is 50 percent greater than in iteration 1. This increase results in a 40 percent rise in the rate of deterioration in 1's trade balance, as the stronger substitution effect leads 1-residents to shift away from the 1-variety to the 2-variety by a greater percentage for each percent decline in the relative price of the 2-variety.

5. Conclusion

This chapter sought to derive a model of bilateral international trade in a non-homogeneous new commodity. Prior to presenting the model, it was indicated that the "Posner-Vernon-Hufbauer" result of low-wage trade supplanting technological gap trade via imitation, with external trade always being monodirectional, can be derived from a simple model which assumed that all new commodities were homogeneous.

Having presented analysis resulting in an expression of the dynamic behavior of the innovating country's trade balance in the ith new non-homogeneous commodity, we submitted the model to computer simulation. With the resulting observations we tested the hypothesis that the model could be approximated by a log linear specification. While this hypothesis could perhaps have been tested by examination of the reduced form equations directly, the exercise allowed us to test the sensitivity of the model to variations in the parameters which

Table 4.2

Growth Rates of the Bilateral Trade Balance:
Log Linear Regression of the Bilateral Trade Balance against Time

Iteration	R_{91}*	R_{92}*	R^2	D.W. (d.f.)
1	3.095 (.0107)	-.045 (.0004)	.997	1.08 (48)
2	3.095 (.0107)	-.085 (.0004)	.999	1.08 (48)
3	3.093 (.0113)	-.055 (.0004)	.998	1.05 (48)
4	3.416 (.0111)	-.088 (.0004)	.999	1.08 (48)
5	2.590 (.0101)	-.034 (.0004)	.995	1.09 (48)
6	3.094 (0108)	-.025 (.0004)	.989	1.03 (48)
7	4.025 (.0014)	-.096 (.001)	.989	1.03 (48)
8	1.683 (.0200)	-.031 (.0007)	.977	1.05 (48)
9	2.701 (.008)	-.027 (.0007)	.995	1.28 (48)
10	2.703 (.008)	-.035 (.0007)	.975	1.21 (48)
11	2.788 (.006)	-.040 (.0002)	.999	1.09 (48)
12	2.989 (.006)	-.043 (.0002)	.999	1.10 (48)
13	3.095 (.010)	-.070 (.0004)	.997	1.08 (48)
14	4.222 (.012)	-.055 (.0004)	.997	1.08 (48)

*The number in parentheses is the standard error of the estimate

Table 4.2

Growth Rates of the Bilateral Trade Balance:
Log Linear Regression of the Bilateral Trade Balance against Time

Iteration	R_{91}*	R_{92}*	R^2	D.W. (d.f.)
15	5.027 (.013)	-.062 (.0004)	.998	1.08 (48)
16	6.271 (.014)	-.072 (.0005)	.998	1.06 (48)
17	4.964 (.011)	-.063 (.0004)	.998	1.08 (48)
18	4.90 (.009)	-.064 (.0003)	.999	1.09 (48)
19	5.23 (.042)	-.048 (.0015)	.960	2.00 (48)
20	5.06 (.013)	-.059 (.0004)	.997	1.08 (48)
21	5.33 (.013)	-.032 (.0004)	.991	1.08 (48)
22	5.33 (.013)	-.082 (.0004)	.999	1.08 (48)
23	5.03 (.014)	-.162 (.0005)	.999	1.26 (48)
24	5.33 (.013)	-.082 (.0005)	.999	1.09 (48)
25	5.34 (.014)	-.082 (.0005)	.998	1.08 (48)
26	5.33 (.013)	-.082 (.0004)	.999	1.09 (48)
27	5.58 (.021)	-.086 (.0007)	.997	1.05 (48)
28	5.13 (.012)	-.113 (.0003)	.995	1.04 (48)

*The number in parentheses is the standard error of the estimate.

could not have been easily accomplished without the regression analysis.

It was ascertained that the innovating country's bilateral trade balance, within the context of a two-country model is a monotone decreasing function of time. In addition, the export (technological gap) flow from innovating to imitating country does not cease to exist for at least fifty time periods for all reasonable values of the parameters of the model. Lastly, we determined that this model can be approximated by a log linear specification.

The model appears to be particularly sensitive to variations in national income growth rates and in income elasticities of demand. It appeared to be less sensitive to variations in the price elasticities, rates of learning, and imitation lags; and not at all sensitive to variations in the initial stock of learning.

FOOTNOTES

[1]On the assumption of everywhere identical factor endowments. Posner, op. cit.

[2]Low-wage exports have displaced technological gap exports according to Hufbauer. Hufbauer, op. cit.

[3]Larger values for c indicate either a more complex new commodity or a less technologically sophisticated imitator; in either case acting to increase the length of the imitation lag OT_L in Figure 4.1.

[4]Balassa, op. cit., p. 89

[5]See Chapter Five, Section 3 for a defense of this assumption. That GDP/cap is a good predictor of the skill endowment of a country is shown by Hufbauer's findings that the Spearman coefficient between skilled employees as a percent of the labor force and GDP/cap was .892. Hufbauer, op. cit., in p. 166.

[6]See Chapter Three, section 5.

[7]Houthakker, H.S. and S.P. Magee, "Income and Price Elasticities in World Trade," Review of Economics and Statistics, May, 1969.

[8]Ibid., Table 4, pp. 166-117.

[9]We confirmed that the latter was the case by simulating values of $\log M_{2t}$ in (4.9) for each of the iterations. In no case did the export flow from 1 to 2 go to zero.

[10]See also, Hufbauer, op. cit.

FIVE

Statistical Analysis

1. Introduction

Implicit throughout the preceding chapters has been the assertion that the product cycle model, as we have developed it, is operational; that is, that it is capable of explaining certain real world phenomena. It is appropos then in this concluding chapter to present evidence, albeit very tentative, supportive of this position.

This chapter is divided into five sections. Section 2 discusses the independent variables to be employed in subsequent empirical analysis. We had hoped to employ very precise measures of three national characteristics which the product cycle suggests should have an impact on the pattern of external trade in new commodities: market size, technological development, and the rate of growth in each. Unfortunately we were continually frustrated in this effort, and as such have employed in our statistical analysis what might, most properly, be termed only very crude proxies of these national characteristics.

Therefore the results of the statistical analysis, which are reported beginning with Section 3, should be interpreted as tentative findings based on rather crude tests of the hypotheses posited therein.

In Section 3, we test the hypothesis that in the chemical sector, the necessary and sufficient conditions are satisfied, so that, when considering a set of potential imitator countries, ranked by relative technological development, a pecking order obtains such that the length of time over which the innovating country has a pair-wise cost advantage is progressively longer.

Applying the tools developed in preceding chapters, Section 4 tests Hirschman's hypothesis that there has occurred an acceleration in the rate of industrial migration.

The simulation study included in Chapter Four suggested that (4.13) could be approximated by a log-linear specification, with time as the independent variable. Section 5 tests the ability of this specification to explain the dynamic behavior of the United States" commodity-country-specific bilateral balance of trade. This test is carried out for seven commodities, and up to fourteen countries -- depending on the particular commodity.

Section 6 investigates a set of hypotheses pertaining to the impact of national characteristics of trading partners of the United States, on the United States' bilateral trade balance in the ith new commodity. Hypotheses set in both cross-sectional and time-series contexts are tested.

Our conclusions are generally positive. Based on our rough tests of each hypothesis, there is tentative evidence supporting the position that, for our sample of commodities and countries and time period, the predictions of the product cycle model agree with real-world experience. Our results, taken together with the growing body of empirical testing of the product cycle model, suggest that Vernon's approach may indeed prove to be a fruitful avenue for including consideration of the technological factor in the received body of comparative cost theory.

2. Independent Variables

Our theoretical analysis suggests four independent variables, each of which should have a predictable impact on either the duration of imitation lags, or on the U.S. ijth bilateral trade balance, the: market size, technological development (the skill endowment), and growth rates of market size and technological development, of any potential imitator. A discussion of the expected impact, à priori, and the measure to be employed for each follows.

Consideration of the influence of each of these attributes on the U.S. cross-sectional, bilateral trade balance is taken by examining their impact on imitation lags. Implicit throughout the subsequent analysis is the assertion that, when considering countries having differing imitation lags in the ith new commodity, the U.S. bilateral trade balance at time t, B_{ijt}, varies directly with the duration of the ijth imitation lag. This assertion follows directly from the bilateral trade model of Chapter Four. Differentiating (4.13) with respect to the imitation lag gives:

$$\partial B_{ijt}/\partial L = d(t+L+e')^{-2}>0. \qquad (5.1)$$

B_{ijt}, in a cross-sectional context, varies directly with imitation lags, and therefore inversely (directly) with cross-sectional variations in national characteristics resulting in shorter (longer) imitation lags in the ith new commodity.

Market Size. Keesing has argued that small countries are at a disadvantage in the international trade of manufactured goods due to limitations imposed by their small home markets, which restrict the exploitation of potential internal and external economies.[1] After correcting Keesing's "fallacy of composition," Balassa has shown that the size effect does indeed play a role in determining intercountry differences in trade patterns, particularly when considering finished manufactured commodities.[2]

We offer evidence below, taken from the experience of the plastics industry, that domestic market size tends to vary inversely with the length of a country's imitation lag in a new commodity. Extending this line of reasoning one step further, to the extent that scale economies, (internal, external, static, or dynamic) are significant, the U.S. trade balance in the ith new commodity at any time t, should deteriorate across an array of countries ranked by (increasing) market size. This is due to the impact of the latter on the duration of the jth country's imitation lag in the ith new commodity.

In order to test this hypothesis, a proxy for market size is required. Our proxy is the jth country's gross domestic product in year t, G_{jt}. We recognize that his measure has drawbacks. It is too broad a definition in the first place. We would have been much happier with a narrower definition, say total sales in the chemical sector when analyzing the behavior of plastics. Unfortunately, the lack of availability of the requisite data precluded use of a more precise measure.

Stobaugh has argued that this measure has additional shortcomings. "The problem in using gross domestic product is its failure to show for some countries, that a large number of people have very low incomes."[3] Stobaugh goes on to suggest total manufacturing output as a superior proxy for market size.

But based on a test using Maizel"s estimates of net value of manufacturing production in billions of 1955 dollars, M_{jt}, and our gross domestic product data, we are unable statistically, to distinguish between these alternative measures of market size.[4]

Simple and Spearman correlation coefficients between M_{jt} and G_{jt} were calculated in a cross-sectional context for the years 1950, 1955, and 1957. The results are presented in Table 5.1. Note the extremely high colinearity between the two alternative measures for all three years; all coefficients are greater than .9. All coefficients are significantly different from zero at the 99% confidence level.

Table 5.1

Correlation Coefficients between Two Measures of Market Size

Year	1950	1955	1957
Simple[a]	.918 (11)	.949 (17)	.949 (11)
Spearman[a]	.927 (11)	.961 (17)	.918 (11)

[a]The number in parentheses below each coefficient is the number of observations.

It would seem, that for our sample of countries, the problem noted by Stobaugh is not serious. There is virtually no difference in the ranking of countries by either measure.

We recognize that even with a relationship as close as that between M_{jt} and G_{jt}, it is not a matter of indifference which measure is used. Two additional considerations led us to choose G_{jt} over M_{jt}.

The collinearity between total manufacturing output and per capita gross domestic product, another measure to be used in our empirical analysis, is significantly greater than between G_{jt} and per capita gross domestic product. The simple correlation coefficients between M_{jt} and per capita GDP are .24, .21, and .23 for the three years; those between G_{jt} and per capita GDP were .17 (1952), .16, and -.01.

But the telling consideration again was the availability of data. The latter was readily available for G_{jt}, and there was unfortunately, a serious lack of same for M_{jt}.

Taking all of the above considerations together, we decided that G_{jt} was the "best" available measure, and therefore chose it as our proxy for market size.

Technological Development. Chapter Two suggests that the duration between innovation and the switching of the relative production cost advantage to the jth potential imitator, varies inversely with the degree of technological sophistication enjoyed by the latter, if the necessary and sufficient conditions were met. It follows therefore, that the same qualitative relationship should obtain between imitation lags and technological development: the greater the latter, the shorter the former. Therefore, if we were to rank U.S. trading partners in the ith new commodity by (increasing) technological sophistication, at time t, we would expect B_{ijt} to deteriorate over this ranking, that is, choosing, for the reasons given below, the jth per capita gross domestic product at time t, g_{jt}, as a proxy for the jth trading partner's technological development, our hypothesis is that $dB_{ijt}/dg_{jt}<0$.

We assumed that a potential imitating country's per capita gross domestic product at time t is a proxy for her relative skill endowment. The choice of this proxy is supported by both Johnson and Hufbauer. Johnson observed that by empirical assumption, the process of economic development entails an upward trend in the value of the original human factor, labor time, as well as generally rising standards of living; consequently, differing values of labor time will obtain between countries at different stages of economic development.[5] Johnson's position is that gross domestic product per capita is a good proxy for relative factor prices between the skill input and ordinary labor input over an array of countries.

Hufbauer has demonstrated for a sample of twenty-four countries in 1965, that per capita gross domestic product is a good predictor of the skill endowment of a country. The Spearman rank correlation coefficient between skilled employees as a percent of the labor force and per capita gross domestic product was .892, and the simple correlation coefficient between these two variables was .879.[6] Both of these coefficients were significant at the 90 percent level.

This proxy is not free from qualifications. It has been suggested that per capita gross domestic product does not adequately proxy supply condition differentials, but rather, that it is a proxy for variations in demand conditions. And associated with this is the argument that a tenet of the product cycle theory is that variations in demand conditions subordinate variations in supply conditions as a determinant of the date of imitation of a new commodity, by a given country.

Demand conditions do play a primary role in determining the initial production location. Indeed, Vernon argues that the locus of initial production is strongly influenced by per capita income level differentials: "Wherever there was a chance to offer a new product responsive to wants at high levels of income, this chance would presumably first be apparent to someone in a position to observe the United States' market. . . Our hypothesis asserts that U.S. producers are likely to be the first to spy an opportunity for high income (new) products."[7]

But when considering the determinants of the direction of imitation (and therefore production) internationally, relative production costs become the primary locational force. "As the demand for a product expands, a certain degree of standardization usually takes place. . . (T)he reduction of uncertainties surrounding the operation enhances the usefulness of cost projections and increases the attention devoted to cost."[8]

But demand conditions according to the above argument, are still important determinants of the route taken by imitation. In fact Vernon's position is not inconsistent with the position that first imitation will occur in economies having a demand structure, as defined by per capita incomes, which is similar to that of the innovator.

The argument that g_{jt} is a good proxy for skill endowment differentials, and that which posits that supply conditions are the sole determinants of migration routes, is weakened by the above considerations. But to the extent that g_{jt} is reflective of both supply conditions, and demand conditions as well, the results presented below are consistent with both the demand and the supply tenets of Vernon's theory.

While we are fully cognizant of the shortcomings of per capita gross domestic product -- it is at best only a rough proxy for technological development, and of the route of migration of comparative advantage in a given new commodity -- it was the best measure for which data was available.

The Rate of Growth. Market size and technological development are not constant over time. The rates at which these characteristics are changing (growing) at any point in time will have an impact on imitation lags.

For example in a world where the number of commodities equals the number of factors, the Rybszynski theorem suggests that growing technological development, entailing as it does an absolute increase in the skill endowment of a country, will, at constant relative world prices, lead to an absolute increase in the optimal production of skill intensive commodities. As skill resources become less rationed relatively, there will obtain an increase in expected profitability from investment in new commodities.

Viewed from the point of view of the potential imitator, such dynamic forces tend to shorten the duration between innovation and switching of relative cost advantage. Thus a potential imitator will seek a license, or endeavor towards successful imitation via independent research at an earlier point in time than would have been the case in an economy in the stationary state.

Considerations of growing market size are also important here. Such growth may result in earlier attainment of the "critical mass" required for successful commercial production in the domestic market of a potential imitator. The impact here will be channeled through the expected revenue flow to be derived from the sale of the ith new commodity. This can be seen by differentiating (4.8), the potential imitator's import time path, with respect to the rate of growth in national income, which gives:

$$d\log M_{2t}/dv_2 = bt>0;$$

where b is the elasticity of demand for the innovating country's variety of the ith new commodity with respect to variations in national income. Therefore the size of the market for the ith new commodity at any time t, is positively related to the rate of growth of national income.

Combining both sets of considerations, we would expect as an immediate result, an inverse relation between the rate of growth in technological sophistication, and/or market size of the jth potential imitator, and her imitation lag in the ith new commodity.

In order to test this hypothesis, we had hoped to include measures of both the rate of growth in technological development, and the rate of growth in the market size. But in our preliminary regression runs, it became clear that the multi-colinearity between these two measures cross-sectionally is severe. Depending upon the year chosen, the partial correlation coefficient between the rate of growth in per capita gross domestic product, and that of gross domestic product ranges between .79 and .93. Countries with rapidly growing market sizes are also countries with rapid rates of growth of technological development.

Since these same preliminary regressions demonstrated that scale economy considerations were not of premier importance, while technological development considerations were, we determined to delete the measure of the rate of growth of market size from the model used in Section 6. We are left then with the rate of change in g_{jt}, r_{jt}, as our measure of the relative state of dynamism in the jth country at time t, and we would expect that B_{ijt} would vary inversely with r_{jt} when viewed in a cross-sectional context.

3. The Duration of the Innovating Country's Cost Advantage

We ascertained in Chapter Two that $N_{W_mV_m} \varepsilon (-1,0)$ was both necessary and sufficient to generate a positive relationship between the ranking of the innovating country's relative skill endowment vis à vis an array of potential imitators, and the ranking of the duration of her cost advantage vis à vis each in any new commodity. Ideally, in order to test the hypothesis $N_{W_mV_m}$ is in the required interval for a given commodity, one would require information on relative returns to skill and to ordinary labor; in addition data on the time lapse between innovation and the switching point of cost advantage for each country in the array of potential imitators would be required. Lacking these ideal measures, we elected to test this hypothesis indirectly via proxy measures for both relative skill endowment, and the elapsed time until switching obtains.

So as to be able to rank countries according to when switching obtains in a given commodity, we assumed that, ceteris paribus, the duration of the innovating country's cost advantage via à vis a potential imitator varies directly with the latter's imitation lag, L_{ij}. This imitation lag served as our proxy for the elapsed time between innovation and switching.

Using as proxies for technological development and market size, those outlined above, the hypothesis can be tested by attempting to predict variations in the duration of imitation lags among countries in a given commodity via variations in national attributes. This test was completed for eleven commodities using the formulation given in (5.2).[9] L_{ij} is the jth country's imitation lag in the ith commodity; G_j is the jth country's gross domestic product in the year that the product was innovated, in billions of 1955 dollars; g_j is the jth country's per capita gross domestic product in the year the ith commodity was innovated, in 1955 dollars;[10] A_{1i} and A_{2i} are the elasticities of the imitation lag with respect to variations in gross domestic product and per capita gross domestic product respectively, for the ith commodity:

$$L_{ij} = A(G_j)^{A_{1i}}(g_j)^{A_{2i}}. \tag{5.2}$$

Market size was included as a variable here so as to net out all influences on the duration of the imitation lag other than the relative skill advantage of the innovating country as measured by per capita gross domestic product. Size effects are of particular concern to the plastics industry from which the sample of eleven commodities was taken. The (plastics) industry is "capital intensive, which ensures that plant size is a direct determinant of operating costs. Whenever there is a large domestic market . . . , it will pay to build larger units, and thereby reduce operating costs."[11]

One would expect therefore, that market size would vary inversely with the duration of the imitation lag in any new commodity, if scale effects were indeed important determinants of unit costs, since a large domestic market is likely to have a negative impact on the duration of the imitation lag.

Before presenting the results, one observation concerning the income data must be made. Maizels' estimates have been used for our purposes here.[12] These estimates are not annual; rather they have been made for five years (of "normalcy") in the first half of the twentieth century: 1899, 1913, 1929, 1937, 1950. For our statistical analysis the innovation date of each commodity was paired with the closest estimate. These pairings appear in Appendix 5.1, Table 5.1.1.

In analyzing our results, we chose to stratify our sample into eight commodities innovated in peacetime and three commodities innovated during World War II. Our rationale was that the processes we examine here can only be expected to work smoothly during peacetime. Therefore we would expect the results for the wartime commodities to be significantly poorer than the peacetime ones.

Looking first at the closeness of the fit, as measured by the $\bar{R}^2$,[13] we see that in five of the eight peacetime commodities the $\bar{R}^2$ exceeds .60 (See Table 5.2). Of the remaining peacetime commodities, in two instances the $\bar{R}^2$ is .49, and only in one does there appear not to be a good fit: polyvinyl acetate.

The fit of the wartime commodities is significantly poorer, the highest $\bar{R}^2$ is .46, and the lowest is .10.

Using a 10% confidence interval as a standard, the regression coefficients of market size are significant in seven of eleven trials, and all estimated coefficients have the proper sign save fluoroethylene. The peacetime commodities again perform much better than the wartime ones: six of the significant coefficients being accounted for by peacetime commodities; and the only improper sign occurring in the case of a wartime good, fluoroethylene.

As expected, scale economies prove to be quite important as a locational force in the plastics industry. On average a doubling of the size of the market as measured by gross domestic product results in a 66 percent decline in the duration of the imitation lag.

Similar results are obtained for the regression coefficient of the sophistication variable. They are significant in five cases, with four of the five significant coefficients occurring with peacetime commodities. All estimated coefficients save acrylonitrile/butadiene/styrene are of the correct sign. These results, then support acceptance -- for commodities innovated in peacetime -- of the hypothesis that for this sample from the plastics industry $N_{W_m V_m}$ is in the interval (-1,0).

Table 5.2

Imitation Lags
(Cross-section Data)[a]

Commodity[b]	constant	A_1	A_2	R^2	D.F.
Celluloid 1870	5.101' (1.90)	-.690' (.23)	-.131 (.36)	.77	4
Phenol Formaldehyde 1909	7.744' (2.22)	-.889' (.24)	-.640' (.38)	.76	7
Alkyd 1926	-1.327 (1.92)	-.841' (.15)	-.754' (.32)	.79	9
Polyester (I) 1942	3.847' (.55)	-.490' (.17)	-.211' (.08)	.43	10
Polyvinyl Acetate 1928	4.918' (1.63)	-.200 (.19)	-.282' (.19)	.182	9
Acrylonitrile/Butadiene/Styrene 1946	-26.964' (13.9)	-3.465 (3.09)	6.267 (3.88)	.95	1
Nylon 1941	4.235 (3.58)	-.218 (.43)	-.302 (.60)	.10	7
Fluoroethylene 1943	9.140 (6.94)	.690 (2.32)	-1.468 (2.21)	.46	2
Nylon Filament 1938	1.862 (1.28)	-.551' (.163)	-.245 (.27)	.49	3
Polyester (II) 1949	5.361' (1.34)	-.560' (.35)	-.328' (.18)	.61	5
Polyproplylene 1949	3.484' (1.55)	-.432' (.21)	-.035 (.24)	.49	5

[a]See Appendix 2 of this Chapter for Data Sources.

[b]Innovation dates follow each commodity.

' Significant at the 90% confidence level.

Note: The number in parentheses below each coefficient is the standard error of the estimate.

4. Acceleration in the Rate of Industrial Migration?

It was demonstrated in Chapter Two, Section 3, that within the context of the model presented therein, the duration of the innovating country's cost advantage varies inversely with the velocity of the product cycle. Therefore during a period when this velocity is accelerating, when comparing successive commodities innovated by a country over time, one would, ceteris paribus, expect the duration of the cost advantage vis à vis each potential imitator, to be decreasing function of time.[14]

It has been suggested that the situation described above reflects the twentieth century experience of the United States. Hirschman feels that

> . . . The recent difficulties of U.S. exports may derive from the fact that we can no longer rely on innovations to throw off a continuous flow of exportables. . . A hypothesis about structural change that could account for our balance of payments difficulties has to do with the speed of industrial migration. . . one has the distinct impression of a shortening in the time necessary for a new commodity to become 'footloose.'[15]

Herein lies a testable hypothesis: if for a given array of countries, the rate of migration of technology is accelerating, when considering a set of commodities ordered chronologically, the imitation lag between innovation and any imitator for each commodity, considered across this ordering, should be a decreasing function of time.

By way of example, suppose that during some time period, a country innovates three commodities, and there exist three imitating countries. If during this time period imitation has progressed in a manner akin to that of Table 5.3, then we would conclude that there has been an acceleration in the rate of industrial migration.

For each of the three countries, the imitation lag for each successive product declines. In such a situation the innovating country becomes progressively less able to rely on long imitation lags to protect her export flows. This Hirschman suggests is the phenomenon faced by the U.S.

Table 5.3

Hypothetical Imitation Lags
(in years)

Commodities listed chronologically	Imitators 1	2	2
1	5	8	11
2	4	7	10
3	3	4	6

Acceleration of industrial migration can be viewed alternatively by considering a special case. Let us define as the "initial imitation lag," the elapsed time between innovation and first foreign production of a commodity. Then, assuming a constant initial imitation lag (as has been approximately the case in the plastics industry.[16]), acceleration of industrial migration can be interpreted as a decline over time in the range of the imitation lags. This can be seen by an appeal to Figure 5.1, which depicts two commodities, w and r, innovated by the same country at two different points in time, with good w having been innovated first.[17] Distance OL^o is the initial imitation lag, which is the same for both w and r by assumption, and L_w^m and L_r^m are the longest imitation lags obtaining for the same array of countries for commodities w and r respectively.

$L_r^m < L_w^m$ reflecting the existence of acceleration. As in Table 5.3, country 3's imitation lag is shorter for good 3 then it was for good 1. Therefore $L^oL_w^m$ and $L^oL_r^m$ measure the range of imitation lags obtaining for each commodity, and, with a constant initial imitation lag, acceleration of industrial migration reduces the range of imitation lags: $L^oL_r^m < L^oL_w^m$.

Continuing with Figure 5.1, c and c' measure the (log transformation) of the highest, and a and a' the (log transformation) of the lowest per capita gross domestic product across the array of imitators on the date each commodity was innovated. Note that c'>c, and a'>a. This reflects the economic growth occurring in the interval between the innovation of w and r: maximal as well as minimal real per capita gross domestic product over the set of potential imitators will be greater at a later point in time.

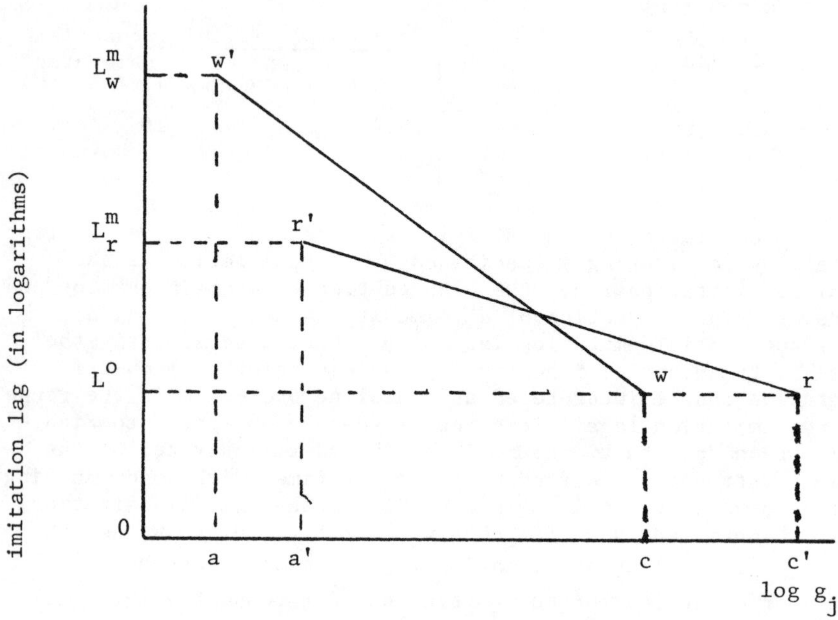

Figure 5.1

Imitation Lags and Per Capita GDP

Given the double-logarithmic form of Figure 5.1, the slope of w'w and r'r measures the elasticity of the imitation lag with respect to variations in g_j, for commodities w and r respectively; or

$A_{2i} = \mathrm{dlog}L_{ij}/\mathrm{dlog}g_j$, and $A_{2w} > A_{2r}$.

If we were to observe declining responsiveness over time of imitation lag duration to variation in either technological sophistication, or country size, cross-sectionally, this analysis would lead us to suspect that there has been an acceleration in the velocity of industrial migration. The above analysis also suggests that such a decline in sensitivity would evidence itself in a decline over time in the elasticity of the imitation lag with respect to variations in either relative per capita gross domestic product, or gross domestic product, A_{1i} and A_{2i} in (5.2)[18].

Evidence for acceleration in the velocity of industrial migration in the plastics industry can be sought by using the estimated values of A_{2i} and A_{1i}. Ranking the commodities chronologically, by date of innovation, and considering only those coefficients which are significant, evidences a downward trend in the absolute value of both elasticities throughout the period (see Table 5.4). The Spearman rank correlation coefficients between the chronological ranking and the ranking by the (absolute) value of A_{1i} and A_{2i} are -.72 and -.60 respectively. Therefore on the basis of this test, it would appear that for the plastics industry, the rate of industrial migration has been accelerating. Once the initial imitation occurs, (usually by Germany, England or Italy), successful commercial production is realized by other potential imitators within increasingly shorter intervals.

Using 5.3, we can perform an, albeit crude, test for acceleration in the rate of migration in the plastics industry using ordinary least squares:

$$\log A_{1i} = B_o + B_1 \log I_i + u_i;$$
$$\log A_{2i} = B_2 + B_3 \log I_i + u_i'. \tag{5.3}$$

A_{1i} and A_{2i} are as defined in (5.2); I_i is the innovation date of the ith commodity; B_1 and B_3 are the elasticities of A_{1i} and A_{2i} (respectively) with respect to variation in innovation dates; and u_1 and u_1' are error terms. Equations (5.3) therefore relate elasticities of imitation lags with respect to variations in G_j and g_j respectively, to the innovation date of the ith commodity.

Using as inputs the observations appearing in Table 5.4, the coefficients of (5.3) were estimated; the results appear in Table 5.5.

Table 5.4

Imitation Lag Elasticities, Selected Synthetic Materials, Ordered Chronologically

Product	Innovation Date	A_{1i}	A_{2i}
Celluloid	1870	-.690	
Phenol Formaldehyde	1909	-.889	-.640
Alkyd	1926	-.841	-.754
Polyvinyl Acetate	1926		-.282
Nylon Filament	1938	-.551	
Polyester (I)	1942	-.490	-.211
Polyester (II)	1947	-.560	-.328
Polypropylene	1949	-.432	

Table 5.5

Results: A Test for Acceleration in the Rate of Industrial Migration

Sample	1870-1949		1900-1949			
	B_o	B_1	B_o	B_1	B_2	B_3
Coefficients	8.36*	-1.11*	25.06*	-3.32*	35.69*	-4.73*
t-value	1.67	1.68	4.27	4.27	1.61	1.61
R^2	.234		.776		.285	
D.F.	5		4		3	

*Significant at the 90% level.

Two estimates of B_1 were made. The first includes a Nineteenth Century innovation, celluloid, innovated in 1870. The second includes only Twenthieth Century innovations. The explanatory power of (5.3) is far stronger when considering only Twentieth Century innovations. The fit as measured by R^2 improves significantly, and both coefficients increase from being less than two times their standard errors to being more than four times their standard error.

Innovation dates do less well in explaining the time profile of the skill elasticity of the imitation lag. The fit as measured by R^2 is quite low. Even so, both coefficients are significant.

An inference to be drawn from these results is that there appears to have been an acceleration of the velocity of industrial migration in the plastics industry during the Twentieth Century: both B_1 and B_3 are significant and less than zero.

5. Growth Rates: the ijth Trade Balance

The simulation study of Chapter 4 indicated that (4.13) could be approximated by a log-linear specification. To be tested in this section is the ability of such an approximation to explain the dynamic behavior of the United States' bilateral trade balance in each of seven commodities for trade with up to fourteen countries. The time period over which the tests are to be run is 1952 to 1965.

The Commodities. Four of the seven commodities, standard typewriters, T.V. tubes and parts, standard electric typewriters, and refrigerators, are considered -- for the sample time period -- as goods whose dynamic bilateral trade balance our model should be able to explain: new commodities in early stages of production.

The remaining three commodities, vacuum cleaners, electric ranges, and radio-phonograph combinations will serve as controls. These are older more-standardized manufactured goods whose external trade behavior our model should not be able to explain.[19]

The Hypotheses. The hypotheses to be tested are that the growth rate of the U.S. balance of trade in the ith new commodity with the jth country, R_{ij2}, should be significantly less than zero, that its absolute value should vary inversely with the level of technological sophistication, and directly with the rate of growth in the per capita income of the jth trading partner.

We tested whether the growth rate of the ijth trade balance is significantly less than zero, using (5.4). B_{ijt} is the ith trade balance between the U.S. and country j at time t; R_{ij2} is the growth rate of B_{ijt} over the sample period; $\hat{R}_{ij1}$ is a constant term; and u_{ij} is an error term.[20]

$$B_{ijt} = X_{ijt}/M_{ijt} = R_{ij1}e^{R_{ij2}t},$$

or:

$$\log B_{ijt} = \log X_{ijt} - \log M_{ijt} = \hat{R}_{ij1} + R_{ij2}(t) + u_{ij}. \quad (5.4)$$

where: $\hat{R}_{ij1} = \log R_{ij1}$.

Estimates of $\hat{R}_{ij1}$ and R_{ij2} were made for each bilateral trade flow using ordinary least squares. These estimates, along with their standard errors, the coefficients of determination, and the degrees of freedom, are presented in Tables 5.6 and 5.12.

"New" commodities. Looking first at the results for the "new" commodities, Tables 5.6-5.9, we note that, although the R^2's are low in many cases, the estimates of the growth rates favor acceptance of the hypothesis under test. The results for each commodity are summarized below, followed by analysis of the Japanese and West German results.

Standard non-electric typewriters. Eleven estimates of the trade-balance growth rate were made for standard non-electric typewriters. Ten of the estimates are significantly different from zero; eight of these significant coefficients have the proper sign.

T.V. tubes and parts. There were ten estimates of A_{ij2} made for T.V. tubes and parts. Seven of the ten estimates were less than zero; six of these were significant at the 90% level of confidence. Only the estimate for West Germany was both significant, and greater than zero.[21]

Standard electric typewriters. As was the case with standard non-electric typewriters, eleven estimates of A_{ij2} were made for electric typewriters. Ten of the eleven are less than zero, and seven are significantly so.

Refrigerators. In the case of refrigerators, twelve estimates of the trade-balance growth rate were made. Nine of the estimates are negative, and eight of these are significantly at the 90% confidence level. All significant coefficients were negative.

Japan. Of the countries sampled, the behavior of the United States' bilateral trade balance with Japan in each of the "new" commodities, provides by far the best, and most consistent support for the hypothesis under test. In all four cases the fit as measured by R^2 is good. The estimates of the trade-balance growth rate are all less than zero, and at least five times their standard errors.

Table 5.6

Growth Rates: jth Trade Balance

Commodity: Standard Typewriters

Country	R_{ij1}[a]	R_{ij2}[a]	R^2	D.F.
Canada	-3.42*	.299	.618	13
	(.077)	(.068)		
Mexico	8.03*	-.358*	.562	35
	(.98)	(.14)		
Sweden	3.14*	-.555*	.488	13
	(1.39)	(.164)		
U.K.	-3.025*	-.1060*	.117	12
	(.7090)	(.08)		
Netherlands	1.000	-.626*	.553	13
	(1.38)	(.162)		
Belgium	23.02*	-3.02*	.516	2
	(12.62)	(2.04)		
France	-1.65	.082	.021	7
	(1.95)	(.196)		
W. Germany	-4.855*	.106*	.162	12
	(.591)	(.069)		
Switzerland	-.409	-.291*	.562	12
	(.633)	(.074)		
Italy	-1.919*	-.354*	.312	12
	(1.2)	(.152)		
Japan	8.262*	-.840*	.822	9

[a]The numbers in parentheses are the standard errors.

*Significant at the 90% confidence level.

Note: Only countries for which data were available appear in Tables 5.6 to 5.12.

Table 5.7

Growth Rates: jth Trade Balance

Commodity: T.V. Tubes and Parts

Country	R_{ij1} [a]	R_{ij2} [a]	R^2	D.F.
Canada	8.542* (1.31)	-.6242* (.132)	.737	8
Mexico	3.832* (.751	.156 (109)	.671	1
U.K.	3.702* (1.379)	-.455* (.139)	.572	8
Netherlands	8.80* (.171)	-.813* (.160)	.811	6
Belgium	7.17* (2.40)	-.072 (.225)	.050	2
France	5.67* (1.77)	-.304* (.178	.266	8
W. Germany	.210 (.572)	.324* (.058)	.794	8
Switzerland	-.351 (3.810)	.390 (.354)	.168	7
Italy	17.606* (7.640)	-1.30* (.641)	.580	3
Japan	12.241*	-1.047*	.777	7

[a] The numbers in parentheses are the standard errors.

*Significant at the 90% confidence level.

Table 5.8

Growth Rates: jth Trade Balance

Commodity: Electric Typewriters

Country	R_{ij1}[a]	R_{ij2}[a]	R^2	D.F.
Canada	6.147*	-.330*	.273	7
	(2.70)	(.200)		
Sweden	5.884*	-.463*	.696	7
	(1.53)	(.116)		
Denmark	4.78	-.145	.132	2
	(3.558)	(.262)		
U.K.	12.718*	-.679*	.689	6
	(2.514)	(.186)		
Netherlands	13.005*	-.800*	.443	7
	(4.491)	(.333)		
Belgium	6.183*	-.119	.149	2
	(2.959)	(.202)		
France	2.492	-.030	.007	1
	(2.002)	(.147)		
W. Germany	4.318*	-.431*	.698	7
	(1.416)	(.107)		
Switzerland	1.596*	-.073	.160	3
	(.834)	(.063)		
Italy	14.885*	-.996*	.447	7
Japan	34.730*	-2.126*	.879	7
	(3.945)	(.298)		

[a]The numbers in parentheses are the standard errors.

*Significant at the 90% confidence level.

Table 5.9

Growth Rates: jth Trade Balance

Commodity: Refrigerators

Country	R_{ij1} [a]	R_{ij2} [a]	R^2	D.F.
Canada	7.688*	-.460*	.422	6
	(2.693)	(.203		
Mexico	.395	-.520*	.540	6
	(2.486)	(.196)		
Sweden	20.058*	-1.502*	.479	6
	(7.840)	(.592)		
Norway	11.436*	-.667*	.936	6
	(1.502)	(.124)		
Denmark	-2.136	.273	.090	1
	(4.751)	(.379)		
U.K.	-1.283	.1070	.020	7
	(3.800)	(.287)		
Netherlands	1.318	.155	.038	5
	(1.796)	(.161)		
Belgium	2.830	-.102	.019	6
	(3.794)	(.295)		
France	15.005*	-.804*	.732	5
	(1.288)	(.126)		
W. Germany	3.963*	-.311*	.295	6
	(2.415)	(.182)		
Italy	10.922*	-.938*	.892	6
	(1.635)	(.123)		
Japan	10.221*	-.750*	.649	6
	(2.765)	(.209)		

[a] The numbers in parentheses are the standard errors.

* Significant at the 90% confidence level.

Table 5.10

Growth Rates: jth Trade Balance

Commodity: Electric Ranges

Country	R_{ij1} [a]	R_{ij2} [a]	R^2	D.F.
Canada	6.246*	.591*	.870	10
	(.533)	(.072)		
Norway	-2.319	.0654	.005	6
	(3.144)	(.375)		
U.K.	-2.754	.564*	.438	7
	(1.938)	(.242)		
W. Germany	.987	-.023	.004	7
	(3.332)	(.422)		
Switzerland	-.507	.037	.001	5
	(3.147)	(.497)		
Spain	1.090	.697	.270	5
	(5.734)	(.662)		
Italy	7.650	-.994*	.503	3
	(5.759)	(.560)		
Japan	18.992*	-1.639*	.608	4
	(6.350)	(.658)		

[a] The numbers in parentheses are the standard errors.

*Significant at the 90% confidence level.

Table 5.11

Growth Rates: jth Trade Balance

Commodity: Radio-Phono Combinations

Country	R_{ij1} [a]	R_{ij2} [a]	R^2	D.F.
Canada	-7.800*	.647*	.350	7
	(4.417)	(.333)		
Mexico	-7.488*	1.185*	.966	2
	(2.198)	(.156)		
Norway	-11.719*	.492	.180	7
	(5.251)	(.396)		
U.K.	-10.517*	1.092*	.730	6
	(3.741)	(.297)		
Netherlands	-12.436*	.330	.09	7
	(5.263)	(.397)		
France	.764	-.216	.010	5
	(12.408)	(.951)		
W. Germany	-22.285*	1.360	.840	7
	(2.971)	(.224)		
Italy	1.408	.179	.448	2
	(1.882	(.142)		
Japan	1.454	-.786*	.238	7
	(7.050)	(.532)		

[a]The numbers in parentheses are the standard errors.

*Significant at the 90% confidence level.

Table 5.12

Growth Rates: jth Trade Balance

Commodity: Vacuum Cleaners

Country	R_{ij1} [a]	R_{ij2} [a]	R^2	D.F.
Canada	15.904*	-.763*	.958	3
	(1.129)	(.113)		
Sweden	-6.817*	.803	.842	2
	(1.728)	(.848)		
U.K.	-9.688*	.729	.313	8
	(2.834)	(.882)		
Netherlands	-6.328*	.447*	.477	9
	(1.257)	(.166)		
France	-8.297	1.114	.567	2
	(6.965)	(.688)		
W. Germany	-9.040*	.88*	.422	10
	(2.432)	(.329)		
Italy	-2.155	.164	.034	3
	(8,085)	(.889)		
Japan	-1.256	-.233	.10	9
	(2.316)	(.275)		

[a] The numbers in parentheses are the standard errors.

* Significant at the 90% confidence level.

These results attest to Japan's oft-mentioned capacity to absorb and imitate foreign technology, and to then exploit successfully her relative ordinary labor advantage (in the sense of Chapter One) vis à vis more developed economies, the result being relatively lower production costs. As an example, in her first four years of exporting T.V. picture tubes, Japan's total exports jumped from eleven units (valued at $283) in 1956, to 2,237 units (valued at $36,884) in 1959. And during the same period, the unit value of T.V. picture tubes from Japan dropped from $25.70 to $16.40. This represents a 36 percent drop in unit values over a four-year span.

Germany. The results for Germany also call for special mention. As with Japan, all four estimates of the trade-balance growth rate are significantly different from zero, but two of the four are positive: standard non-electric typewriters and T.V. tubes and parts.

We can shed some light on these seemingly contradictory results by appealing to a paper by Gruber, Mehta and Vernon.[22] They demonstrated therein that the export profile of West Germany was very much like that of the United States. This is due, they asserted, to the fact that West Germany, like the United States is on top of the advanced country list, having a relatively high per capita gross domestic product, and heavy stress on industrial innovation and product development; Germany therefore, derives her export strength from roughly the same characteristics as those that govern the United States' export performance.[23]

Control Commodities. Looking now at the control group of commodities, we observe that our model is, as expected, generally not able to explain the dynamic behavior of these bilateral trade balances. For the three commodities taken together, twenty-five estimates of the trade-balance growth rates were made. Of these only eleven are significant at the 90% confidence level.

It is interesting to note that seven of the significant coefficients are positive. A possible explanation for this is that American corporations have introduced successive (relatively skilled intensive) generations of these commodities, which are successfully replacing the older ones in international markets. For example, during this period -- 1952 to 1965 -- the vacuum-tube radio-phonograph has been to a large extent, replaced in international markets by transistorized sets, the production of which American corporations has a dominant position. By exploiting her relative skill advantage, the United States is attempting, successfully, to regain a portion of her earlier share in each of these markets by specializing in the new, relatively skill-intensive generations of these commodities.

The Japanese results in the control group warrant a bit of discussion. The product cycle model asserts that countries with relatively greater skill-disadvantages (in the sense of Chapter One), will have longer imitation lags since the switching point, T*, takes longer after innovation to obtain. Therefore such a country's external trade in a new commodity should exhibit the behavior predicted by our model -- a negative trade-balance growth rate in this case -- (long) after the innovating country is no longer able to exploit its relative skill advantage vis à vis technologically more developed countries. We would therefore, predict significant negative growth rates in the United States bilateral trade balance with Japan in each of the control commodities.

In all three cases the estimates of R_{ij2} for Japan are less than zero. They are significantly different from zero in two of the three cases: electric ranges and radio-phonograph combinations. This result lends further credence to the validity of our model.

The relative magnitudes of R_{ij2}. It was determined, via the simulation study, that the magnitude of the trade-balance growth rate was quite sensitive to variations in the rate of growth of per capita income; in fact almost a 1:1 relationship was discerned. Secondly, it was shown therein that the absolute value of this growth rate varied directly with the length of the imitation lag.

We have also shown earlier -- in Chapter One -- that the length of the imitation lag can be expected to vary inversely with relative technological development, a proxy for which is gross domestic product per capita.

These two hypotheses were tested using the Spearman rank correlation coefficient. Each country whose estimated trade-balance growth rate was significantly less than zero was first ranked by the absolute value of its trade-balance growth rate.

This ranking was set sequentially against those by 1952 GDP/capita, 1965 GDP/capita, and the 1952-1965 average growth rate in per capita gross domestic product. Spearman rank correlation coefficients were then calculated.

The results appear in Table 5.13, and while they cannot be considered decisive, they do favor acceptance of both hypotheses. Save for refrigerators, the rate of deterioration of the U.S. bilateral trade balance does vary inversely with per capita gross domestic product. In all cases this rate of deterioration varies directly with the rate of growth of per capita income.

<u>Summary</u>. We have, in this section, tested three hypotheses generated from the theoretical analysis of previous chapters: the growth rate of the United States' bilateral trade balance in the ith new commodity appears to be significantly different from zero, and negative. Further, it is very tempting to conclude that the magnitude of this growth rate will vary inversely with the technological sophistication, and directly with the rate of growth in per capita gross domestic product of the United States' jth trading partner.

Table 5.13

Spearman Rank Correlation Coefficients between the Ranking by the Absolute Value of R_{ij2} and that by:

Commodity	1952 GDP/capita	1965 GDP/capita	1952-1965 d(GDP/cap)/dt
Standard Typewriters	-.20	-.35	.60
T.V. Tubes & Parts	-.57	-.80	.35
Electric Typewriters	-.80	-.89	.65
Refrigerators	.15*	.20*	.24*

*Not significant at the 90% level.

6. National Characteristics and the ijth Trade Balance

This section investigates hypotheses pertaining to the impact of national characteristics of United States' trading partners on the former's bilateral trade balance in the ith new commodity. The preceding analysis (Chapters One to Four) suggests three attributes of a trading partner as potentially having significant influence on the United States' bilateral trade balance: market size, degree of technological development, and rate of growth. The relative importance, and the interpretation of the impact of each characteristic are tested in a cross-sectional and a pooled-data context. Both investigations are carried out with the same sample of commodities over the same time period as that of Section 5 of this chapter; four of the commodities are deemed new commodities, in early stages of production during the sample time period, and three of the commodities, considered as mature commodities, serve as a control group.

Cross-sectional Analysis. One of the premier goals of our theoretical analysis is to be able to explain the cross-sectional variation in the U.S. bilateral trade balance in a new commodity. That is, when considering an array of U.S. trading partners at time t, the U.S. bilateral trade balance will vary in a systematic fashion -- the direction of which our theory can predict -- with respect to variations over this array in the three attributes cited above.

We have then three variables, each of which should have significant influence on the cross-sectional behavior of B_{ijt}. For purposes of testing for the hypothesized impact of each national attribute on B_{ijt}, using least squares techniques, the three national attributes were combined into a single-equation econometric model (5.5). In (5.5) a_{it}, d_{it}, and k_{it} are the partial elasticities of B_{ijt} with respect to variations in G_{jt}, g_{jt}, and r_{jt}, respectively; u_{ijt} is an error term; c is the constant term of the regression equation. All elasticities are expected to take on negative values.

$$\log B_{ijt} = c + a_{it}\log G_{jt} + d_{it}\log g_{jt} + k_{it}\log r_{jt} + u_{ijt} \tag{5.5}$$

Two sets of cross-sectional regressions were run on (5.5). The first set was run by dividing the period 1950-1967 into six three-year sub-periods. Three-year averages of each bilateral trade flow, as well as three-year averages of G_{jt} and g_{jt} were calculated.

For the growth variable we elected to employ five-year average growth rates as calculated by the Agency for Internationaal Development. We felt that these would best reflect the state of growth expectations in each sub-period. The principle problem with this procedure is that it reduced greatly the cross-sectional variation of this variable, and therefore its explanatory power relative to that of the three-year moving average growth rates used in the subsequent annual cross-sectional regressions.

The coefficients of these regressions along with their standard errors -- in parentheses below each estimate -- the R^2's, and the degrees of freedom, are presented in Tables 5.18-5.24 (at the end of the chapter).

A second set of regressions was run on (5.5). Annual cross-sectional data was employed except in the case of the growth variable for which three-year moving averages of the rate of growth of per capita gross domestic product were employed.[24] Regressions were run for each year for which a sufficient number of observations was available.

The choice of the three-year moving average growth rate was based in large part, upon the belief that the three-year moving average presents a better picture of the dynamic state of an economy at a given point in time, than do either annual growth rates, or two-year moving averages, since both the latter two can be expected to be plagued with short-run "noise".

When compared against the five-year average growth rate, we felt that the three-year moving average could be expected to be more sensitive to varying conditions of dynamism across an array of countries within a cross-sectional context of a single year.

A subsidiary, but important consideration in this choice was the fact that in the course of the preliminary tests of the model, none of the annual, two-year moving average, or the five-year average growth rates performed nearly as well as the three-year moving average. In addition, the three-year moving average presented us with much less multicolinearity than did any of the other candidates.

For the above reasons, we report below only the results obtained when the three-year moving average growth rate was included. The estimates of the coefficients in the annual cross-section regressions runs, along with other pertinant results appear in Tables 5.25-5.31.

New Commodities. The performance of the model within the three-year average and annual contexts is summarized in Tables 5.14 and 5.15 respectively. A general inference to be drawn from these is that there is evidence for product cycle phenomena as enunicated by Vernon, and as developed above.

Looking first at the R^2's, we note that in eight of fourteen three-year average runs, and in sixteen of twenty-eight annual runs, R^2's of .50 or better were obtained, with much better fits on average resulting from the annual cross-sectional runs. This would suggest that our three independent variables do have a significant impact on intercountry variations in B_{ijt}.

The estimated coefficients are consistently significant at the 90 percent confidence level. In the three-year average runs the elasticity coefficients of market size, sophistication and the growth variable, are significant in 71%, 86%, and 59% of the estimates respectively; all significant coefficients are of the proper sign. Similarly in the annual runs, these same coefficients are significant in 54%, 72% and 82% of the estimates; all of the significant estimates are less than zero, save the estimate of k_{i1962} for standard non-electric typewriters.

Technological development, as measured by g_{jt}, proves to have a significant as well as consistant influence on the behavior of the U.S. bilateral trade balance. In addition to being significant in 86% and 72% of the two sets of runs respectively, the values taken by d_{it} are consistently large: depending on the commodity and the three-year sub-period chosen, the U.S. can expect that a one percent increase in per capita gross domestic product, cross-sectionally, will result in from a 2 percent to an 8.2 percent deterioration in its bilateral trade balance. Of the four commodities, bilateral trade in T.V. tubes and parts exhibits the greatest sensitivity to cross-sectional variations in per capita gross domestic product: d_{it} takes on an average value of -6.5, and varies between -5.3 and -8.2 depending upon the sub-period chosen. In the annual regression runs, the range of d_{it} is -2.1 to -8.7 with a mean of -5.2.

Both Vernon and Hufbauer suggest that one important determinant of the locus of innovation, and the speed and direction of imitation, is variation in market size across an array of countries.[25] This would suggest that B_{ijt} should be sensitive to variations in market size. But for our sample of commodities, and for the time period chosen, B_{ijt} does not appear to terribly sensitive to cross-sectional variations in G_{jt}. Save for the 1965-1967 sub-period estimate for refrigerators, where a_{it}, the is -4.59, the absolute value of three-year estimates of a_{it} is consistently less than 2.0. In fact, in four of the ten significant estimates of a_{it}, the resulting value is inelastic, a one percent increase in market size led to less than a one percent deterioration in the U.S. trade balance.[26]

As evidenced in Table 5.16, whether the results are based on three-year average or the annual data are chosen, the U.S. bilateral trade balance is consistently and uniformly more sensitive to variations in skill endowment than variations in market size. This would suggest that while there is good evidence supporting the product cycle on both counts -- variation in technological sophistication as well as in market size -- technological considerations and not scale economy ones, are of particular importance in explaining external trade in our four "new" industries.

Table 5.14

Summary
Three-year Cross-section Regressions

Commodities	Total Estimates	Significant			Proportion Significant		
		a_{it}	d_{it}	k_{it}	a_{it}	d_{it}	k_{it}
Std. Non-electric Typewriters	5	3	3	2	.60	.60	.40
T.V. Tubes & Parts	3	3	3	1	100	100	.33
Std. Electric Typewriters	3	1	3	3	.33	100	100
Refrigerators	3	3	3	1	100	100	.33
Total	14	10	12	7	.71	.86	.50

Table 5.15

Summary
Annual Cross-section Regressions

Commodities	Total Estimates	Significant			Proportion Significant		
		a_{it}	d_{it}	k_{it}	a_{it}	d_{it}	k_{it}
Std. Non-electric Typewriters	12	7	7	9	.59	.59	.75
T.V. Tubes & Parts	4	2	2	4	.50	.50	100
Std. Electric Typewriters	5	3	4	4	.60	.80	.80
Refrigerators	7	3	7	6	.45	100	.86
Total	28	15	20	23	.54	.72	.82

Table 5.16

Average Elasticity Values

Commodity	Three-year Average			Annual Average		
	a_{it}	d_{it}	k_{it}	a_{it}	d_{it}	k_{it}
Standard Non-electric Typewriters	-1.3	-4.7	-1.5	-2.2	-4.9	-2.6
T.V. Tubes & Parts	-1.2	-6.5	-1.4	-1.3	-7.6	-3.0
Standard Electric Typewriters	-0.3	-3.7	-1.8	-8.9	-5.1	-2.9
Refrigerators	-2.2	-3.8	-3.9	-0.9	-3.2	-3.8
Total	-1.3	-4.7	-2.9	-1.4	-5.2	-3.1

Control Commodities. As expected (5.5) is incapable of consistently explaining the cross-sectional behavior of B_{ijt} in either context, when considering the control commodities. In the case of the three-year average runs, a total of eleven estimates of each elasticity were made. None of the estimates of a_{it} were significantly different from zero at the 90 percent confidence level. Three of the estimates of d_{it} were significant; one of them being greater than zero.

Three estimates of k_{it} were significant; all three were estimates for radio-phonograph combinations. One of the significant estimates is positive. In fact, the U.S. bilateral trade balance in radio-phonographs appears quite sensitive cross-sectionally to variations in per capita gross domestic product. As suggested before, in section 5 of this chapter, this sensitivity could be due in part to the new wave of imitation set off by the U.S. innovation of transistorized radio-phonographs. Evidence for this position comes from the realization that by the mid-sixties, Japan had begun producing, and exporting transistorized sets back to the United States.

Turning to the annual regressions, the model retains its loss of empirical validity when considering other than new commodities in early stages of production. There were, for the combined group of control commodities, sixteen estimates made of each elasticity. Of these two estimates of a_{it} were significant, and one of the two was positive. There were five significant estimates of d_{it}, four of which were of the proper sign. For k_{it} six of the sixteen estimates were significant, and five were of the proper sign.

It should be noted here that the lion's share of these significant estimates -- all but four -- are accounted for by radio-phonograph combinations in the sixties -- which we have suggested above might be a special case -- and the remainder are due to that in vacuum cleaners. Not a single significant estimate is realized for trade in electric ranges.

Pooled Data Analysis. We had hoped to test our model, (5.5) within a time-series context, but we were confronted with extremely high multicolinearity between gross domestic product and per capita gross domestic product for every country in our sample (See Appendix 5.2, Table 5.2.6). The traditional remedy for this problem, taking first differences, did not help. Due to a lack of population growth of significant proportions in most countries in our sample, between 1950 and 1968, the multicolinearity between the first difference array of these two variables for each country was also unacceptable (See Appendix 5.2, Table 5.2.8). Therefore we concluded that either of these approaches would fail to isolate the separate contributions of G_{jt} and g_{jt} to variation in B_{ijt}.

Klein suggests that since "econometric information is scarce and that any fresh samples providing an independent insight should not be over looked."[27] He subsequently demonstrates that situations which involve such technical problems as multicolinearity, can be dealt with by pooling the cross-section and time-series samples.[28]

Following Klein's lead, we elected to pool the samples for each of the seven commodities. Using these new samples, we ran seven more regressions with the specification given by (5.5). The estimated coefficients, along with their standard errors (in parentheses below each), the R^2, and degrees of freedom for each commodity are presented in Table 5.17.

We note first that the fit as measured by R^2 is generally weak. This can in part be attributed to the inclusion of the isolated observations for countries not included in cross-sectional regressions, i.e., Ireland, Portugal, and Spain are examples. The result was that for each commodity, the variance of the sample of trade balances increased, with the result that the goodness of fit of the regression plane decreased.

In terms of significance from zero, and signs of the estimated coefficients, the model again appears to agree with real-world phenomena. For the four new commodities, every estimate of each elasticity is significantly less than zero. For the control commodities on the other hand, only three of nine total estimates were significant; all three were less than zero. Therefore our model remains capable of distinguishing between new and mature commodities.

Variations in technological development again have the strongest relative impact on the U.S. bilateral trade balance. Both on average, and in each commodity equation, d_i takes on the largest absolute value.

Table 5.17

Pooled Regressions: Bilateral Trade Balances and National Characteristics (pooled data, 1950, 1968)[a]

Commodity	c	a_i	d_i	k_i	R^2	D.F.
NEW:						
Std. Non-electric Typewriters	29.104* (4.75	-.775* (.33)	-3.439* (.56)	-1.219* (.53)	.356	126
T.V. Tubes & Parts	15.541* (5.58)	-.940* (.33)	-1.005* (.67)	- .842* (.49)	.29	68
Std. Electric Typewriters	17.027* (6.77)	-.654* (.35)	-2.389* (.83)	- .207* (.11)	.334	46
Refrigerators	20.149* (5.12)	-.938* (.34)	-1.602* (.62)	-1.514* (.81)	.313	55
Average Elasticity		-.826	-2.111	- .950		
CONTROL:						
Vacuum Cleaners	-35.04* (11.76)	.141 (.46)	-4.528* (4.39)	1.208 (1.34)	.24	52
Radio-Phonograph Combinations	25.262* (12.56)	1.09 (.98)	-3.549* (1.5)	-.058 (1.63)	.36	28
Electric Ranges	3.740 (10.08)	.073 (.28)	- .187 (1.21)	-1.37* (.70)	.31	53

*Significant at 90% confidence level.

[a]See Appendix 5.2 for data sources.

[b]Simple average of the elasticity coefficients significantly less than zero.

On the average the trade balance is more sensitive to variations in growth rates than in market size. But in two commodities, T.V. tubes and parts and electric typewriters, the trade balance proves to be more sensitive to market size variations. This result is at variance with those of Table 5.15 (annual cross-section runs). In our annual cross-section analysis, market size was invariable the least important influence on the vehavior of the bilateral trade balance.

Summary. This portion of this section, investigates hypotheses pertaining to the impact of cross-sectional variation in national attributes of U.S. trading partners, on the U.S. bilateral trade balance in the ith new commodity.

We began our investigation by suggesting that B_{ijt}, at time t, varies directly with cross-sectional variation in imitation lags. We then examined the potential impact of these national attributes on the length of imitation lags in the ith new commodity. This exercise allowed us to ascertain the potential impact of variation in these attributes on B_{ijt} within a cross-sectional context. The result was a one-equation econometric model, (5.15).

The results of the tests completed with this model suggest that the theory has empirical validity. That is to say, real-world phenomena failed to disprove the hypotheses being tested.[29]

For the two cross-sectional, and the pooled-data studies, virtually all significant coefficients were of the expected sign, and a respectable proportion of the estimates were significantly different from zero when considering the four "new" commodities.

We performed the same exercise using three control commodities, whose trade behavior our model should not be able to explain. It could not do so. Based on this exercise, we tentatively suggest that our model appears able to discriminate between new, relatively skill-intensive commodities, and older, more mature commodities.

Lastly, our statistical analysis suggested that the U.S. bilateral trade balance, for our commodity sample, is on average, quite sensitive to cross-sectional variation in technological sophistication (as the product cycle theory would suggest), somewhat, but less sensitive to the rate of growth in per capita gross domestic product; and only slightly sensitive to variation in market size.

Table 5.18

Three-year Average Trade Balances
(Cross-sectional Data, 1950-1967)[a]

Commodity: Standard Non-Electric Typewriters

Period	c	a_i	d_i	k_i	R^2	D.F.
1950-1952	52.834* (19.2)	-1.431* (.68)	-5.237* (2.06)	-2.325* (1.20)	.565	6
1953-1955	34.223 (32.54)	-1.118 (1.13)	-5.003* (3.42)	- .648* (.147)	.331	5
1956-1958	31.86* (15.55)	-1.939* (1.11)	-3.949* (2.18)	-1.299 (1.10)	.408	9
1962-1961	-7.625 (16.95)	- .454* (.174)	.911 (2.31)	.819 (3.04)	.246	7
1965-1967	-1.15 (12.95)	.099 (.98)	- .431 (.976)	-3.319 (3.77)	.210	9

[a]See Appendix B of this chapter for data sources.

*Significant.

Note: Only time-periods for which data were available appear in Tables 5.18 to 5.24.

Table 5.19

Three-year Average Trade Balance (Cross-sectional Data, 1950-1967)[a]

Commodity: T.V. Tubes and Parts

Period	c	a_i	d_i	k_i	R^2	D.F.
1956-1958	45.03*	-1.837*	-5.317*	-.812	.536	4
	(24.79)	(1.04)	(3.00)	(1.40)		
1959-1961	41.88*	- .1567*	-6.251*	4.065	.4744	5
	(22.75)	(.058)	(3.29)	(4.71)		
1962-1964	50.66*	-1.367*	-8.199*	-1.380*	.614	5
	(24.70)	(1.01)	(4.18)	(.851)		

[a]See Appendix B of this chapter for data sources.

*Significant.

Table 5.20

Three-year Average Trade Balance[a]
(Cross-sectional Data, 1950-1967)

Commodity: Standard Electric Typewriters

Period	c	a_i	d_i	k_i	R^2	D.F.
1959-1961	42.726*	- .318*	-6.140*	-.795	.522	5
	(19.9)	(0.13)	(2.68)	(.351)		
1962-1964	21.797	- .023	-2.625*	-3.078*	.530	6
	(18.384)	(.06)	(1.55)	(1.65)		
1965-1967	15.280*	1.01	-2.258*	-1.498*	.610	8
	(10.28)	(.98)	(1.00)	(1.00)		

[a]See Appendix B of this chapter for data sources.

*Significant

Table 5.21

Three-year Average Trade Balances[a]
(Cross-sectional Data, 1950-1967)

Commodity: Refrigerators

Period	c	a_i	d_i	k_i	R^2	D.F.
1959-1961	19.185*	- .738*	-2.257*	-.691	.367	11
	(10.09)	(.60)	(1.34)	(1.94)		
1962-1964	24.120*	-1.234*	-2.070*	-3.945*	.484	6
	(11.22)	(1.01)	(1.07)	(2.39)		
1965-1967	-67.198*	-4.586*	-7.175*	.519	.880	3
	(21.93)	(1.03)	(2.94)	(1.47)		

[a]See Appendix B of this chapter for data sources.

*Significant

Table 5.22

Three-year Average Trade Balances[a]
(Cross-sectional Data, 1950-1967)

Commodity: Electric Ranges

Period	c	a_i	d_i	k_i	R^2	D.F.
1950-1952	-26.674	.528	3.742	2.159	.378	2
	(31.736)	(.881)	(4.60)	(2.49)		
1956-1958	41.553*	.139	-6.167*	-2.832	.416	4
	(25.90)	(1.01)	(3.78)	(2.41)		
1959-1961	-9.995	-.435	1.977	.209	.272	4
	(33.71)	(.841)	(4.32)	(4.42)		
1962-1964	-5.881	.580	2.183	1.771	.289	6
	(1.10)	(.00)	(2.95)	(1.89)		

[a] See Appendix B of this chapter for data sources.

* Significant.

Table 5.23

Three-year Average Trade Balances
(Cross-sectional Data, 1950-1967)[a]

Commodity: Radio-phonograph Combinations

Period	c	a_i	d_i	k_i	R^2	D.F.
1959-1961	40.572*	- .687	-5.113*	-6.462*	.582	6
	(15.19)	(1.05)	(2.21)	(2.81)		
1962-1964	43.279	-1.325	-3.671	11.633*	.617	4
	(53.27)	(1.36)	(7.02)	(5.57)		
1965-1967	83.49*	.140	-11.876	-8.015*	.664	4
	(54.65)	(1.15)	(8.79)	(3.33)		

[a]See Appendix B of this chapter for data sources.

*Significant.

Table 5.24

Three-year Average Trade Balances
(Cross-sectional Data, 1950-1967)[a]

Commodity: Vacuum Cleaners

Period	d	a_i	d_i	k_i	R^2	D.F.
1950-1952	-107.45*	.821	15.676*	5.518	.699	2
	(52.905)	(1.87)	(7.32)	(4.46)		
1956-1958	-48.87*	.367	7.06*	3.250	.538	2
	(24.3)	(1.66)	(3.89)	(3.55)		
1959-1961	-68.900	1.863	9.00	4.01	.363	4
	(56.7)	(2.07)	(7.18)	(7.12)		
1962-1964	-40.03	-.986	5.064	2.060	.310	6
	(38.1)	(.963)	(4.39)	(4.85)		

[a]See Appendix B of this chapter for data sources.

*Significant

Table 5.25

Annual Trade Balances
(Cross-sectional Data, 1952-1967)[a]

Commodity: Standard Non-Electric Typewriters

Year	c	a_i	d_i	k_i	R^2	D.F.
1952	34.783*	-4.911*	-4.752*	-2.032*	.663	6
	(17.79)	(2.31)	(2.21)	(1.08)		
1953	44.554*	-2.155*	-4.122*	-2.662*	.641	6
	(14.28)	(1.23)	(1.75)	(1.26)		
1954	5.730	-1.760*	.667	-1.520	.464	4
	(34.166)	(1.07)	(3.94)	(2.76)		
1955	-3.192	-.889*	1.10	-1.484*	.7050	4
	(15.04)	(.51)	(1.73)	(1.01)		
1956	-40.516*	-.509*	-5.275*	-1.860*	.603	4
	(20.46)	(.28)	(2.46)	(1.05)		
1957	48.754*	-.917	-6.015*	-1.802*	.481	8
	(20.90)	(1.73)	(2.72)	(.98)		
1958	41.553*	-1.919*	-4.93*	-1.979*	.432	9
	(19.59)	(1.24)	(2.42)	(1.33)		
1959	45.055*	-.515	-5.669*	-3.051*	.535	7
	(16.604)	(1.38)	(2.17)	(2.17)		
1960	20.451	.779	-3.459*	-2.126*	.227	8
	(16.60)	(1.61)	(2.26)	(2.40)		
1961	-9.102	-.142	-1.318	-3.711*	.326	5
	(8.61)	(.145)	(1.28)	(2.08)		
1962	-26.79	-3.448*	2.146	5.114*	.538	6
	(19.65)	(2.26)	(2.75)	(2.59)		
1965	-8.251	-.456	.687	-5.225	.241	8
	(14.406)	(.85)	(1.94)	(4.58)		

[a]See Appendix B of this chapter for data sources.

*Significant

Note: Only years for which data were available appear in Tables 5.25 to 5.31.

Table 5.26

Annual Trade Balances
(Cross-sectional Data, 1952-1967)[a]

Commodity: T.V. Tubes and Parts

Year	c	a_i	d_i	k_i	R^2	D.F.
1957	16.692*	-.520	-1.236	-.562*	.240	3
	(12.51)	(1.95)	(2.58)	(.29)		
1958	61.223*	-.773	-6.974*	-2.65*	.773	3
	(18.74)	(.91)	(2.40)	(1.08)		
1960	16.495*	-1.399*	-1.033	-2.237*	.551	5
	(12.66)	(1.05)	(1.77)	(1.08)		
1963	-60.630*	-1.266*	-8.189*	-6.642*	.523	6
	(28.39)	(.765)	(3.47)	(3.98)		

[a]See Appendix B of this chapter for data sources.

*Significant

Table 5.27

Annual Trade Balances
(Cross-sectional Data, 1952-1967)[a]

Commodity: Standard Electric Typewriters

Year	c	a_i	d_i	k_i	R^2	D.F.
1961	22.677*	-.318*	-3.638*	-6.425*	.426	4
	(14.84)	(1.96)	(2.02)	(3.54)		
1962	29.509*	-1.078*	-3.754*	-3.564*	.921	3
	(9.08)	(.242)	(1.05)	(2.03)		
1963	68.312	.078	-8.640*	-.416*	.585	7
	(24.42)	(.73)	(2.86)	(.26)		
1964	29.643*	-1.27*	-4.444*	-.756	.637	6
	(20.53)	(.84)	(2.56)	(2.52)		
1965	-23.553	1.902	1.663	-1.337*	.370	5
	(48.63)	(1.30)	(5.66)	(.79)		

[a]See Appendix B of this chapter for data sources.
*Significant

Table 5.28
Annual Trade Balances
(Cross-sectional Data, 1952-1967)[a]

Commodity: Refrigerators

Year	c	a_i	d_i	d_i	R^2	D.F.
1961	24.328*	-.990*	-2.492*	-1.372*	.390	10
	(12.23)	(.65)	(1.41)	(.93)		
1962	16.654	.269	-2.147*	-8.967*	.503	10
	(13.28)	(.51)	(1.53)	(3.99)		
1963	17.639	.380	-2.138*	-1.989*	.5892	9
	(15.51)	(.88)	(1.20)	(.97)		
1964	32.881*	-.714*	-3.043*	-3.653	.487	5
	(14.87)	(.121)	(1.95)	(5.78)		
1965	29.817*	-2.02*	-1.912*	-1.16*	.560	6
	(13.64)	(.94)	(.67)	(.63)		
1966	37.74*	-.280	-3.981*	-3.902*	.410	7
	(17.26)	(1.47)	(2.27)	(2.38)		
1967	45.368*	1.67	-6.488*	-5.866*	.678	4
	(17.87)	(1.93)	(2.50)	(2.43)		

[a]See Appendix B of this chapter for data sources.

*Significant

Table 5.29

Annual Trade Balances
(Cross-sectional Data, 1952-1967)[a]

Commodity: Radio-Phonograph Combinations

Year	c	a_i	d_i	k_i	R^2	D.F.
1960	39.897	-1.119	-3.527	-7.454	.465	2
	(76.99)	(1.76)	(8.53)	(7.76)		
1961	23.958	1.158	-2.967	-9.067*	.436	9
	(18.50)	(.85)	(2.16)	(2.95)		
1963	-4.961	1.209	.877	-7.028*	.605	6
	(14.96)	(1.94)	(1.68)	(2.57)		
1964	31.397*	1.586	-4.540*	-4.115*	.458	7
	(16.64)	(1.65)	(2.12)	(2.07)		
1965	140.632*	.0368	-16.942*	-9.190*	.535	5
	(69.75)	(1.31)	(8.31)	(3.89)		
1966	66.977*	1.641*	-8.832*	-6.372*	.787	6
	(17.46)	(1.01)	(2.07)	(1.84)		

[a]See Appendix B of this chapter for data sources.

*Significant.

Table 5.30

Annual Trade Balances
(Cross-sectional Data, 1952-1967)[a]

Commodity: Vacuum Cleaners

Year	c	a_i	d_i	k_i	R^2	D.F.
1952	-66.712	.391	9.531	.870	.501	2
	(91.42)	(2.13)	(11.38)	(6.07)		
1953	-68.672*	-.527*	-9.607*	2.277*	.992	1
	(7.51)	(.16)	(.98)	(.28)		
1959	-29.156	-.627	4.682	-.709	.685	1
	(104.54)	(9.08)	(17.27)	(22.92)		
1960	-45.924	.878	5.025	2.267	.326	2
	(49.80)	(1.71)	(5.85)	(5.20)		
1963	-57.692*	-.975	5.657*	-21.561	.537	6
	(33.12)	(.97)	(3.57)	(10.68)		

[a]See Appendix B of this chapter for data sources.

*Significant.

Table 5.31

Annual Trade Balances
(Cross-sectional Data, 1952-1967)[a]

Commodity: Electric Ranges

Year	c	a_i	d_i	k_i	R^2	D.F.
1952	-.454	-.433	-.092	-3.651	.365	2
	(28.70)	(.62)	(3.89)	(3.08)		
1959	-1.431	.251	.246	-1.014	.447	2
	(25.35)	(.59)	(3.40)	(1.52)		
1960	6.910	.790	-1.164	-1.919	.264	3
	(40.82)	(1.080)	(5.23)	(2.37)		
1961	39.897	-1.12	3.53	7.45	.36	2
	(76.99)	(1.76)	(8.53)	(7.76)		
1963	-4.487	.642	.637	-2.318	.269	5
	(30.56)	(.78)	(3.40)	(3.68)		

[a]See Appendix B of this chapter for data sources.

*Significant.

FOOTNOTES

[1]Keesing, D.B., "Population and Industrial Development: Some Evidence from Trade Patterns," American Economic Review, June, 1968, pp. 448-455.

[2]Balassa, Bela, "Country Size and Trade Patterns," American Economic Review, March, 1969, pp. 201-204. For an excellent discussion of these and related considerations, see Bela Balassa, Trade Liberalization Among Industrial Countries, Chapter 5.

[3]Stobaugh, "Where in the World Should We Put that Plant?," Harvard Business Review, Jan./Feb., 1969, p. 130.

[4]Maizels' estimates appear in Alfred Maizels, Industrial Growth and World Trade, Appendix B, Table B3.

[5]H.G. Johnson, op. cit., pp. 32, 33.

[6]Hufbauer, op. cit., in Vernon, ed., op. cit., p. 166.

[7]Vernon, op. cit., pp. 192, 194.

[8]Ibid., p. 196.

[9]The double logarithmic transformation of (5.2) was used in the regressions. The choice of this specification was predicted by virtue of its generally better fit and ease of interpretation.

[10]See Appendix 5.2 for data sources.

[11]O.E.C.D., Gaps in Technology: Plastics, p. 56.

[12]A. Maizels, Industrial Growth and World Trade, Appendix B, Table B1 and Table B11.

[13]Adjusted for degrees of freedom.

[14]Note that if the innovating country is at the same time experiencing increasing relative labor disadvantage in the sense of Chapter Two, these forces will be reinforced.

[15]A.O. Hirschman, "An Invitation to Theorizing of the Dollar Glut," Review of Economics and Statistics, May, 1969.

[16]See Appendix 5.1.

[17]A diagram similar to that of Figure 5.1 could be drawn depicting the relationship between imitation lags and gross domestic product. Since both the diagram and the reasoning would virtually reproduce that which follows, we chose not to treat gross domestic product separately. Suffice it here to say that the result will be that of A_{1i} will also decline over time when acceleration in migration obtains.

[18]Formulation 5.2 is very restrictive. Since a speed-up in industrial migration over time could be due to either a convergence of gross domestic produce and per capita gross domestic product between countries, or to a downward shift in the constant term in 5.2, the failure of A_{1i} or A_{2i} to decline over time would not preclude the existence of acceleration. But on the other hand, a negative time trend in each of these elasticities would make it very tempting to conclude the acceleration had obtained in the industry in question.

[19]For a more complete background discussion of each of the seven commodities, see Appendix 5.2.

[20]The rationale for the choice of this functional form follows directly from the results of the analysis of Chapter Four. The equilibrium bilateral trade balance time path is given by (4.13); whose growth rate, (4.14), is everywhere negative for all positive t. The subsequent simulation study of (4.13) demonstrated that this time path could be approximated for by the functional form given by (5.4) -- which is identical to that of (4.15).

[21]A possible rationale for this result is given below.

[22]Gruber, W., Mehta, D., & Vernon, R., _op_. _cit_.

[23]_Ibid_., p. 26.

[24]See Appendix 5.2 for a discussion of the calculation of these moving averages.

[25]Vernon, _op_, _cit_.; Hufbauer, op. cit.

[26]This apparent lack of sensitivity could be due to the crudity of our proxies.

[27]Klein, L., _An Introduction to Econometrics_, p. 62.

[28]_Ibid_., p. 62. See the discussion on pp. 62-79 for other situations in which an appeal to pooling can be beneficial.

[29]We remain mindful here of the tentative quality of these conclusions due to the crudity of our proxy measures.

APPENDICES

Appendix 2.1

The purpose of this appendix is to demonstrate that the assumption of a Cobb-Douglas production function is not necessary to generate the results presented above. The choice of the Cobb-Douglas form was predicated upon a desire to generate the results in succinct, explicit form.

Consider a general production function of the form:

$Q = F(K,L,S,t)$. Introducing dynamic biases in optimal input mix via factor augmentation terms, and defining the inputs in efficiency units gives:

$$Q = F(e^{xt}K, e^{yt}L, e^{zt}S). \quad \text{(A1)}$$

By defining N_K, N_L, N_S as the elasticities of output with respect to changes in capital, ordinary labor, and skill respectively, we have:

$$N_K = (F_K e^{xt})K/F,\ N_L = (F_L e^{yt})L/F, \text{ and } N_s = (F_S e^{zt})S/F; \quad \text{(A2)}$$

which with linear homogeneity and perfect competition also represent the distributive shares for the three factors.

It can be shown that the elasticity of an isoquant for any two inputs taken together is equal to the ratio of their distributive shares; that is: $dL/L/dS/S = N_S/N_L = N_{LS}$.[1] And similarly for capital and skill taken together: $N_{KS} = N_S/N_K$.

Since capital is assumed internationally mobile, the behavior of N_{KS} is not a locus-determining factor.[2] But it will be shown that the initial value as well as the time profile of N_{LS} determines both the location of initial production and dynamic shifts in the locus of cost advantage.

The isoquant elasticity concept is presented diagrammatically in Figure A2.1 for skill and ordinary labor taken together. The elasticity of the isoquant at A is given by:

$$N_{LS} = dL/L/dS/S = (BL_o/L_o)(L_oA/AS_o) = BL_o/AS_o = BA/AC. \quad \text{(A3)}$$

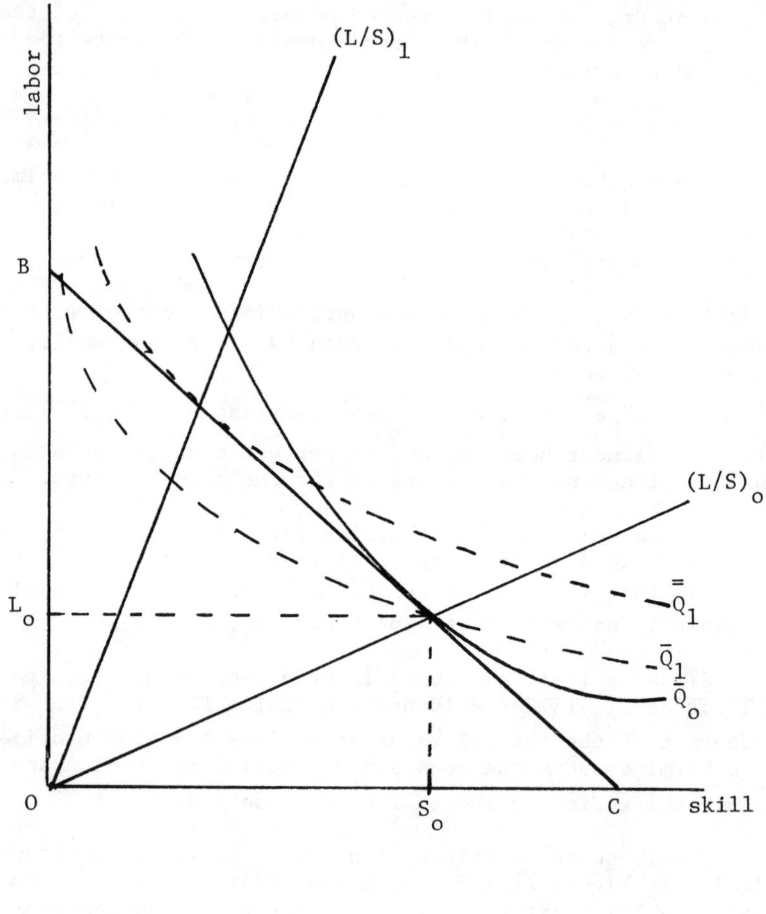

Figure A2.1
Elasticity of an Isoquant

Therefore, $N_{LS} = N_S/N_L = BA/AC$.

The dynamic variations in optimal input mix introduced earlier imply that $dN_{LS}/dt < 0$ for any given ratio, viz., ratio $(L/S)_o$: in order to maintain $(L/S)_o$ on the expansion path over time, BC would have to rotate counter-clockwise shortening BA and lengthening AC, i.e., resulting in a lower value for N_{LS} along the ray $(L/S)_o$. Thus the elasticity of $\bar{Q}_o$ at A declines over time which says that the $\bar{Q}_o$ isoquant rotates in a counter-clockwise direction for $\bar{Q}_o$ to $\bar{Q}_1$ between periods t=0 to t=1, and the optimal factor mix obtains as $(L/S)_1 > (L/S)_o$ with output equal to $\bar{Q}_1 > \bar{Q}_o = \bar{Q}_1$, and $N^1_{LS} < N^2_{LS}$.[3]

It has been shown by Bertrand and Vanek that the necessary and sufficient condition for such variations in optimal input mix to obtain is that $y > z$ obtain.[4] This indicates that requiring dj_t/dt $(=dN_S/dt)$ to be negative above was unnecessarily strong. A much weaker condition will generate the desired results: $dN_L/dt > dN_S/dt$.[5]

The isoquant elasticity concept can be used to indicate how variations in the optimal input mix affect the locus of production cost advantage within the context of a general production relation. Figure A2.2 depicts isocost planes for country 1 (ABC) and country 2 (AB'C'). Since total expenditure can be determined as r(OA), the two planes represent equal levels of expenditure (or total cost).

In which of the two countries the cost advantage obtains for a given commodity at a given point in time is determined by the facit of AC'DB with which the highest isoquant for that commodity is tangent: country 1 for all tangencies is ADB, and country 2 for all tangencies in ADC'. Should the relevant isoquant for some commodity be simultaneously tangent to both facets, neither country has a production cost advantage, and production can take place in both country 1 and country 2 simultaneously.

A necessary but not sufficient condition for country 1 to enjoy initial production of a new commodity is that $N_{LS} > DC/DB$ obtain. This will guarantee that the isoquant for the new commodity is tangent in the ADB facet at the date of innovation.

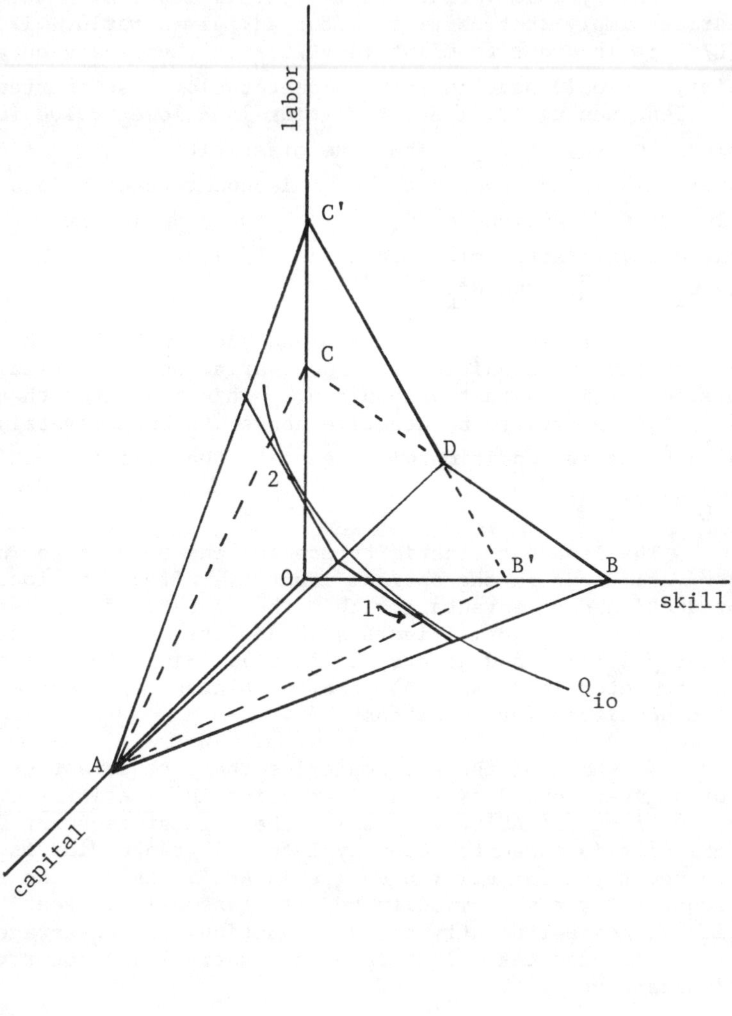

Figure A2.2

Isoquant Elasticity and Variation in the Optimal Input Mix

But it does not rule out the possibility that the same isoquant is simultaneously tangent to the ADC' plane, viz., Q_{io} in Figure A2: N_{LS}^{o} at 1 is greater than DC/DB, but the isoquant is also tangent to ADC' at 2. Thus a necessary (and sufficient) condition is that N_{LS}^{o} be greater than that obtaining at the point of simultaneous tangency. This will guarantee a unique point of tangency in K-L-S space for the relevant isoquant at the date of innovation.

By reverting to the Cobb-Douglas formulation given by (1) above, the necessary and sufficient conditions for a simultaneous tangency (and therefore for country 1 to enjoy initial production) can be obtained explicitly. Recalling that total cost is equal everywhere on AC'DB, and that a simultaneous tangency requires the same output in each country, it follows that:

$$(r/a)^{a}(w_1/b)^{b}(s_i/j)^{j} = (r/a)^{a}(w_2/b)^{b}(s_2/j)^{j} \qquad \text{(A4)}$$

in a situation of simultaneous tangency. By solving (A4) for the ratio j/b, the necessary and sufficient condition for a simultaneous tangency in terms of N_{LS} and relative factor prices can be determined:[6]

$$(j/b)^* = W/-V - N_{LS}^*. \qquad \text{(A5)}$$

Thus the critical value of the elasticity of the isoquant is determined by relative factor prices. For given magnitudes of W and B, country 1 will enjoy initial production of a new commodity if and only if: $N_{LS}^{o} > N_{LS}^{*}$ at the point of tangency of the highest isoquant and AC'DB.[7] And the locus of production cost advantage shifts to country 2 at the point in time, T* when $N_{LS}^{T*} = N_{LS}^{*}$.

FOOTNOTES

[1]This is proved in, J.C.H. Fei and G. Ranis, _Development of the Labor Surplus Economy_, pp. 69-72.

[2]This indicates that the above-assumed constancy of the elasticity of output with respect to changes in capital was an expedient, and not necessary for the results. It is recognized that this assumption was quite restrictive in the sense that in a situation in which both N_S and N_L are varying, it is difficult to expect N_K to remain constant.

[3]Given that the necessary and sufficient conditions for a monotone decreasing marginal cost time path noted above are satisfied.

[4]T.J. Bertrand and J. Vanek, "Growth with Technological Change, Variable Returns to Scale and a General Saving Function," _La Rivista Internazionale de Scienze Economiche e Commerciali_, 1969.

[5]But if both dN_L/dt and dN_S/dt were positive, then dN_K/dt would have to be negative in order to preserve linear homogeneity. As indicated in Chapter Two, the change in distributive shares is a necessary but _not_ sufficient condition for the new technique to be adopted.

[6](A5) also demonstrates that the elasticity of the isoquant at the point of tangency in ADB is equal to that at the point of tangency in ADC' for the simultaneous tangency situation, viz., isoquant Q_{io} in Figure A2: the elasticity of Q_{io} at 1 equals that at 2.

[7]Note that N_{LS}* is greater than DC/DB in Figure A2 since it obtains to the southeast of D along DB.

Appendix 4.1

In this appendix we deduce and prove a proposition derived from the model of Chapter Four.

P.1'. $X(t)$ is maximized at T_L.

Proof: This can be proved geometrically as long as the functions are well-behaved. Consider Figure A4.1.1.

Claim: $X(t)$ for t less than T_L are less than $T_L a'$.

Observe: $X(t) = t \tan\theta$; over the interval OT_L both t and angle Θ are increasing and therefore so is $X(t)$.

Claim: Beyond T_L, $dX(t)/dt<0$.

Consider some later point in time t'. Construct lines bb' and aa' each parallel to 0a'. Then the addition to $X(t)$ over the interval $T_L t'$ is given by the distance a'b, and similarly the subtraction in $X(t)$: $T_L a$. Since $T_L a$ is greater than a'b (and this will hold over the entire interval $T_L T^*$), there has been a net decline in $X(t)$.

<u>Therefore</u>: for all $t>T_L$, $dX(t)/dt<0$; and therefore $X(t)$ is maximized at $t = T_L$.

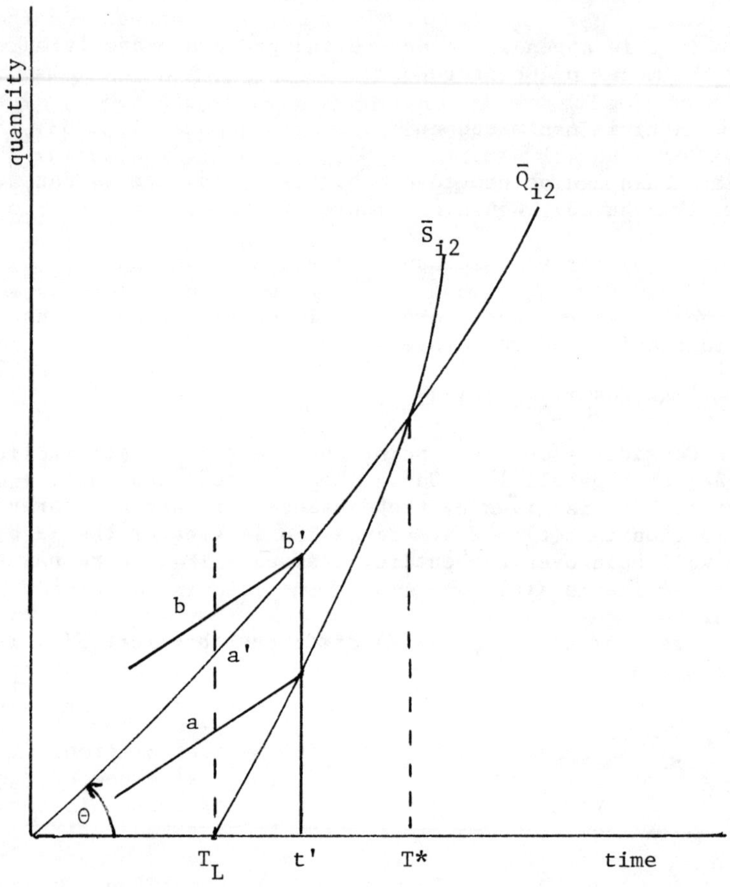

Figure A4.1.1

Maximal Technological Gap Exports in a Homogeneous New Commodity

Appendix 5.1

This appendix presents the data used in the empirical tests of the Theorem derived in Chapter Two. Table A5.4 presents the first production dates of twelve commodities in the plastics industry for fourteen countries. Table A5.1.2 presents estimates of gross domestic product and per-capita gross domestic product for the same countries for the period 1899 - 1957. Table A5.1.3 presents the pairings between innovation dates and innovation dates used in the empirical analysis of Chapter Five.

Table A5.1.1

First Production Dates of Selected Synthetic Materials

Thermoset Plastics

Country	Celluloid	Phenol Formaldehyde	Alkyd	Polyester
United States	1870	1909	1926	1942
Germany	1872	1910	1927?	1953
United Kingdom	1879	1910	1929	1950
France	1875	1916	1928	1950
Italy	1924	1922	1927	1949
Japan	1908	1923	1931	1953
Canada	----	1911	1932	1950
Netherlands	1919-1935	1928	1940	1950
Sweden	----	1917-1954	1942	1952
Belgium	----	1934	1938	----
Switzerland	1923	1937?	1937?	1955?
Norway	----	1950	1950	1955
Denmark	----	1936	----	----
Mexico	----	1851	1951	1957

Thermoplastic Plastics

Country	Acetate	Acrylonitrile /Butadiene/ Styrene	Nylon	Fluoro-ethylene
United States	1928	1946	1941	1943
Germany	1928	1955	1943	1953
United Kingdom	1949	1962	1950	1945
France	1937	1960	1943	1953
Italy	1954	----	1946?	1955
Japan	1936	1963	1953	1963
Canada	1931	----	1957	----
Netherlands	1955	----	1950	----
Sweden	1955	----	----	----
Belguim	----	----	1955?	----
Switzerland	1942	----	1953	----
Norway	1956	----	----	----
Denmark	----	----	----	----
Mexico	1952	----	1956	----

Table A5.1.1
(Continued)

Synthetic Man-Made Fibers

Country	Nylon Filament	Polyester Nylon Staples	Polyproplene Polyester
United States	1938	1949	1949
Germany	1949	1958	1956
United Kingdom	1941	1954	1951
France	1941	1954	----
Italy	1938	1954	----
Japan	1942	1958	1959
Canada	1942	1955	1960
Netherlands	1950	1955	1956
Sweden	1963	----	1960
Belgium	1955	----	----
Switzerland	1950	----	----
Norway	1959	----	1961?
Denmark	----	----	1962
Mexico	1965?	1956	1959

Source: G.C. Hufbauer, _Synthetic Materials and the Theory of International Trade_, Table C-3.

Table 5.1.2

Estimates of Gross Domestic Product and Per Capita Gross Domestic Product for Selected Countries 1899-1957[a]

	1899		1913		1929		1937		1950	
Country	G	g	G	g	G	g	G	g	G	g
United States	59.0	79.0	97.0	100	168	1380	171	1330	294	1940
Germany	29.3	525	37.5	560	40.5	625	30.6[b]	725[b]	31.8[b]	665[b]
United Kingdom	34.0[c]	830[c]	42.0[c]	920[c]	42.0	915	50.0	1055	54.7	1085
France	14.0[f]	360	16.0	400	25.0	605	22.4	540	32.4	775
Italy	6.05[e]	185	8.00	225	11.1	275	11.3	260	16.7	360
Japan	2.8	65	4.8	90	9.1	145	12.9	185	11.1	135
Canada	3.42	635	7.2	975	10.3	1030	10.1	895	20.0	1460
Netherlands	2.2[f]	430	3.08	505	4.37	560	5.2	600	7.55	750

Table 5.1.2
(Continued)

	1899		1913		1929		1937		1950	
Country	G	g	G	g	G	g	G	g	G	g
Sweden	1.83	360	2.42	430	4.07	655	5.25	835	8.3	1185
Belgium	3.32[g]	480	5.75	730	6.05[h]	800	6.87	1150	10.3	1315
Switzerland	1.29	390	1.80	475	2.87	720	2.97	705	4.80	1020
Norway	0.83[f]	375	1.15	480	1.77	630	2.29	720	3.23	980
Denmark	1.03	430	1.72	6.5	2.55	730	3.11	840	4.08	950
Mexico	----	---	----	---	----	---	4.30	230	6.00	235

Source: A. Maizels, Industrial Growth and World Trade.

[a]Gross domestic product (G) in billions of 1955 dollars; Per capita gross domestic product (g) in 1955 dollars.
[b]West Germany excluding West Berlin and the Saar.
[c]British Isles.
[d]1895-1905
[e]1901
[f]1900
[g]1895
[h]1930

Table A5.1.3

Pairs of Innovation Dates and Estimates of National Characteristics

Commodity	Innovation Date	Estimate
Celluloid	1870	1899
Phenol Formaldehyde	1909	1913
Alkyd	1926	1929
Polyester	1942	1938
Polyvinyl Acetate	1928	1929
Acrylonitirle/Butadiene/ Styrene	1946	1950
Nylon	1941	1937
Fluroethylene	1943	1937
Nylon Filament	1938	1937
Polyester Nylon Staple	1949	1950
Polypropylene	1949	1950

Appendix 5.2

Section 1: Country Sample
National Characteristics

Section 2: Commodity Sample

Section 3: Data Sources

1. Country Data

Statistical information on trade with the U.S., and on national economic attributes for the period 1952 to 1968, was collected for seventeen countries, who together account for the lion's share of U.S. imports. The seventeen nations appear in Table A5.2.1, along with the exchange rates per dollar used to translate the data from national currencies to dollar equivalents.

Data on gross national product and per capita gross national product, both in constant 1967 prices, appear in Table A5.2.2 and Table A5.2.3 respectively. Table A5.2.4 presents five-year growth rates of gross national product and per-capita gross national product. These data were all obtained from the A.I.D.'s *Gross National Product, Growth Rates and Trend Data by Region and Country*, Washington, D.C., 1969.

Table A5.2.1

Seventeen Countries Studied -- With Exchange Rates Per Dollar

Canada	1.081 c$
Mexico	12.5 pesos
Sweden	5.17 kroner
Norway	7.14 kroner
Denmark	6.9 kroner
U.K.	.357 pounds
Ireland	.357 pounds
Netherlands	3.62 guilders
Belgium	50 B francs
France	4.937 F francs
Germany, Fed. Rep.[a]	4 D.M.O.
Switzerland	4.3 S francs
Finland	4.2 markkas
Spain	60 pesetas
Portugal	28.75 escudos
Italy	625 lire
Japan	360 yen

[a]plus West Berlin

Table A5.2.2

Gross National Product in Constant 1967 Prices
(Billions of dollar equivalents)

Country	1952	1953	1954	1955	1956	1957	1958	1959	1960
Canada	310	322	313	339	369	374	378	391	400
Mexico	97	98	108	118	125	135	142	146	157
Sweden	130	135	144	148	153	157	161	170	176
Norway	45	47	49	50	53	54	55	56	58
Denmark	65	69	71	71	73	77	79	84	89
U.K.	706	736	766	792	803	819	826	859	901
Ireland	22	23	23	23	23	23	23	24	25
Netherlands	110	120	128	138	142	147	145	152	166
Belgium	112	116	120	126	130	133	132	135	143
France	558	573	599	630	666	700	719	741	796
Germany (W.)	524	566	611	682	729	771	796	852	925
Switzerland	82	88	92	97	103	106	104	112	118
Finland	000[a]	000[a]	000[a]	43	44	45	45	48	53
Spain	125	124	133	140	151	154	162	155	162
Portugal	21	23	24	25	25	27	27	29	30
Italy	229	322	334	356	372	392	411	438	465
Japan	295	316	336	368	398	445	463	509	586

Country	1961	1962	1963	1964	1965	1966	1967	1968
Canada	411	438	461	490	524	558	573	598
Mexico	164	171	181	200	211	226	241	258
Sweden	186	192	202	217	226	234	239	248
Norway	62	65	69	72	76	79	83	86
Denmark	95	100	101	110	115	118	122	124
U.K.	933	945	988	1044	1072	1092	1109	1147
Ireland	26	27	28	29	30	30	32	33
Netherlands	171	178	184	200	211	216	227	239
Belgium	150	158	165	177	184	189	195	203
France	830	889	942	1003	1051	1108	1159	1208
Germany (W.)	975	1015	1050	1119	1183	1210	1210	1293
Switzerland	127	133	140	147	153	157	160	166
Finland	57	59	61	65	68	69	71	72
Spain	181	196	210	225	242	261	269	281
Portugal	32	34	36	39	41	43	46	49
Italy	502	533	562	578	598	632	670	703
Japan	675	726	782	891	924	1020	1157	1295

Source: Agency for International Development, Statics & Reports Division, Office of Program & Policy Coordination, _Gross National Product, Growth Rates and Trend Data, by Region and Country_, Washington, D.C., April 25, 1969.

[a]not available.

Table A5.2.3

Per Capita Gross National Product in Constant 1967 Prices
(dollar equivalents)

Country	1952	1953	1954	1955	1956	1957	1958	1959	1960
Canada	2140	2163	2039	2157	2288	2240	2210	2231	2236
Mexico	337	348	344	354	363	391	385	403	420
Sweden	1832	1880	1989	2103	2150	2136	2170	2280	2348
Norway	1353	1392	1444	1459	1520	1546	1550	1590	1634
Denmark	1485	1559	1595	1585	1611	1681	1717	1821	1921
U.K.	1397	1453	1507	1553	1568	1593	1600	1653	1721
Ireland	745	767	795	799	794	802	796	830	878
Netherlands	1061	1142	1206	1279	1307	1332	1299	1341	1445
Belgium	1285	1323	1364	1424	1458	1482	1461	1485	1558
France	1315	1344	1391	1450	1520	1580	1606	1638	1742
Germany (W.)	1030	11101	1177	1302	1376	1437	1467	1552	1669
Switzerland	1709	1794	1864	1956	2044	2070	2004	2126	2206
Finland	a	a	a	1019	1028	1033	1023	1090	1189
Spain	440	433	461	483	514	523	542	516	536
Portugal	247	264	277	286	295	307	310	325	345
Italy	632	675	697	738	767	804	838	887	938
Japan	345	364	382	413	443	490	506	551	628

Country	1961	1962	1963	1964	1965	1966	1967	1968
Canada	2248	2355	2434	2544	2671	2781	2301	2877
Mexico	451	475	455	485	494	513	528	546
Sweden	2467	2539	2656	2835	2927	2994	3043	3130
Norway	1723	1792	1871	1949	2047	2107	2198	2264
Denmark	2024	2120	2117	2289	2378	2410	2482	2489
U.K.	1767	1771	1940	1932	1970	1995	2014	2071
Ireland	924	950	986	1019	1037	1051	1092	1146
Netherlands	1468	1510	1538	1653	1717	1732	1802	1877
Belgium	1629	1711	1780	1885	1941	1982	2039	2107
France	1799	1892	1968	2071	2148	2243	2324	2399
Germany (W.)	1733	1783	1822	1919	2003	2028	2021	2154
Switzerland	2311	2357	2418	2496	2575	2624	2640	2702
Finland	1274	1316	1336	1410	1472	1497	1522	1547
Spain	593	636	676	718	764	819	838	866
Portugal	361	381	401	425	449	459	489	513
Italy	1055	1061	1110	1130	1159	1217	1279	1358
Japan	718	765	816	920	943	1032	1158	1283

Source: <u>Ibid</u>.

a not available.

Table A5.2.4

Five-Year Average Growth Rates

	Gross National Product				Gross National Product/Capita			
Country	A	B	C	D	A	B	C	D
Canada	4.6	3.4	5.4	3.6	1.8	0.8	3.6	1.7
Mexico	6.2	6.1	5.9	6.8	3.1	3.0	2.8	3.2
Sweden	3.4	3.4	5.1	2.9	2.7	2.8	4.5	2.2
Norway	4.0	3.3	5.4	4.5	3.0	2.4	4.6	3.7
Denmark	2.0	4.7	5.2	2.5	1.2	4.0	4.4	1.6
U.K.	2.7	2.8	3.3	2.5	2.5	2.3	2.5	1.9
Ireland	7.0	5.2	4.0	4.9	6.7	4.9	3.4	4.4
Netherlands	5.6	4.1	4.7	5.3	4.4	2.8	3.3	4.1
Belgium	3.4	2.5	4.4	3.7	2.9	1.9	3.7	3.1
France	4.1	5.0	4.8	4.4	3.3	4.0	3.3	3.4
Germany	9.5	6.3	4.7	3.5	8.6	5.2	3.4	3.0
Switzerland	5.0	3.9	5.2	2.7	3.8	2.4	3.0	1.5
Finland	3.5	4.1	5.2	2.2	2.7	3.2	4.6	1.6
Spain	7.1	3.7	9.3	3.7	6.3	2.9	8.5	2.8
Portugal	3.8	4.7	7.9	6.8	3.3	4.2	7.1	5.1
Italy	6.0	5.8	5.2	5.4	5.4	5.2	4.5	4.6
Japan	9.0	9.5	9.7	12.7	8.3	8.6	8.7	11.5

Source: Ibid.

A: 1950-1954
B: 1955-1959
C: 1960-1964
D: 1965-1969

Three-year Moving Average Growth Rates. In the annual, cross-sectional analysis of Chapter Five, the logarithmic transformation of the three-year moving average of g_{jt} was employed. These moving averages were calculated from the data appearing in Table III of this Appendix. Observations for 1950 and 1951, needed to calculate averages for 1952, and 1953, do not appear in Table III, but they were obtained from the same AID publication cited as the source for those data.

The formula used to calculate these averages was:

$$r_{jt} = (g_{jt+2}/g_{jt-1})^{\frac{1}{2}} - 1.$$

Multicolinearity. In order to test for multicolinearity in the data, simple correlation coefficients were calculated for each of the six possible pairs of the logarithmic transformation of the four potential national attributes to be used in the analysis of Section 6 of Chapter Five. This exercise was run for the data arranged in cross-section, time-series, and pooled arrays. The results for each appear in Tables A5.2.5, A5.2.6, and A5.2.7, respectively.

At issue here is the set of variables which can be included, in the model to be tested, as independent variables. Including two variables which are highly colinear, would result in high covariance among the estimated parameters, and thus it would not be possible to discover the separate influence of the independent variables. On the other hand, exclusion of a plausible independent variable because that variable is highly correlated with the other independent variables imparts a bias to the estimators of the other coefficients. Consider a model:[1]

$$y_t = a_o + a_1 x_{1t} + a_2 x_{2t} + a_3 x_{3t} + u_t;$$

where the a_i's are the coefficients of the independent variable x_{it}, and u_t is an error term. Excluding the variable x_3 due to its colinearity and x_1 and x_2 gives:

$$y_t = a_o + a_1 x_{1t} + a_2 x_{2t} + u_t';$$

where $u_t' = a_3 x_{3t} + u_t$, and will be correlated with x_1 and x_2, and therefore a_1 and a_2 will be biased. As Christ observes, we are faced with the choice between "two 'goods', low correlation among explanatory variables, and inclusion of an explanatory variable if we have some confidence in its relevance. . ."[2]

Looking through Tables A5.2.5, A5.2.6, and A5.2.7, we are seriously confronted with this choice in only two instances. Taken in any of the three contexts, the correlation coefficient between the rate of growth in gross national product and in per capita gross national product is at least .66, and in most cases is over .90. Since the theoretical justification for inclusion of GNP growth rates is marginal, it was determined, on the basis of this test, to exclude this variable from subsequent analysis.

The second choice situation involved the strong colinearity of GNP and GNP/capita in the time series context. The correlation coefficient for all countries, save Germany (0.17), was .97 or greater. Since there was also serious multicolinearity in the time-series array between GNP and the three-year moving average growth rate in GNP.capita, we determined to use first differences rather than un-transformed time-series data. But as can be seen in Table A5.2.8, this ploy did not improve the situation. As there was relatively weak theoretical justification for testing the model derived in Section 6 of Chapter Five in a time-series context, we, on the basis of these tests, elected not to pursue either time-series, or first difference analysis.

Table A5.2.5

Correlation Coefficients: Cross-sectional Context (1952-1968); Log Transformation

Year	G/G-c	G/R-c	G/R	G-c/R-c	G-c/R	R-c/R
1952	.77	.17	.22	-.05	-.004	.90
1953	.77	.17	.22	-.05	-.002	.90
1954	.77	.18	.26	-.03	.002	.90
1955	.26	.16	.19	-.33	-.29	.90
1956	.28	-.03	.12	-.39	-.21	.79
1957	.28	-.01	.14	-.32	-.25	.79
1958	.29	-.003	.15	-.29	-.23	.79
1959	.29	.009	.17	-.24	-.26	.79
1960	.29	.03	.18	-.23	-.25	.79
1961	.29	-.13	-.09	-.28	-.29	.81
1962	.15	.04	-.04	-.30	-.20	.88
1963	.20	.02	-.07	-.26	-.21	.87
1964	.24	-.09	.004	-.21	-.27	.87
1965	.24	-.10	-.001	-.19	-.25	.87
1966	.29	.09	.10	-.27	-.30	.93
1967	.28	.17	.12	-.30	-.32	.93
1968	.29	.13	.14	-.28	-.29	.93

Key
G: Gross National Product at Constant 1967 prices.
G-c: Per Capita Gross National Product at Constant 1967 prices.
R-c: Three-year moving average growth rates in G-c.
R: Five-year growth rates of G.

Table A5.2.6

Correlation Coefficients: Time Series Context

Country	G/C-c	G/R-c	G/R	G-c/R-c	G-c/R	R-c/R
Canada	.97	.33	-.03	.30	-.06	.92
Mexico	.98	-.10	.36	-.16	.32	.88
Sweden	.99	.14	.16	.15	.17	.99
Norway	.999	.63	.58	.64	.59	.998
Denmark	.999	.253	.28	.25	.27	.998
U.K.	.999	-.57	.135	-.56	.14	.69
Ireland	.99	-.67	-.65	-.72	-.70	.997
Netherlands	.999	-.07	-.05	-.07	-.05	.996
Belgium	.999	.48	.53	.50	.54	.996
France	.999	-.21	.26	-.21	.26	.66
Germany	-.17	.23	.10	-.96	-.95	.98
Switzerland	.99	-.69	-.44	-.70	-.45	.86
Finland	.99	.12	.09	.14	.11	.99
Spain	.99	-.10	-.07	-.10	-.07	.99
Portugal	.99	.72	.84	.74	.85	.98
Italy	.99	-.89	-.86	-.90	-.87	.99
Japan	.99	.76	.80	.76	.81	.99

Table A5.2.7

Correlation Coefficients
(Pooled Data, 1952-1968)

	G	G-C	R-C	R
G	1.00	.61	.32	.19
G-C	.61	1.00	-.29	-.34
R-C	.32	-.29	1.00	.91
R	.19	-.34	.91	1.00

Key: G: Gross National Product at Constant 1967 Prices.

G-C: Per-Capita GNP at constant 1967 prices.

R-C: Three-year, moving-average growth rates in G-C.

R: Five-year average growth rates of G.

Table A5.2.8

Correlation Coefficients
(First Differences, 1952-1968)

Country	G/G-C
Canada	.81
Mexico	.20
Sweden	.72
Norway	.94
Denmark	.98
U.K.	.99
Ireland	.74
Netherlands	.48
Belgium	.44
France	.97
Germany	.96
Switzerland	.91
Finland	.56
Spain	.99
Portugal	.57
Italy	.67
Japan	.95

2. Commodity Data

A sample of seven commodities was chosen for the statistical analysis of Sections 5 and 6 of Chapter Five. The data on U.S. external trade flows in these commodities were taken from Department of Commerce sources.[3] Since the Department of Commerce classifies exports and imports on the basis of different schemes, it was necessary to determine that the content of the export (Schedule B) classification was virtually identical to that of the import (Schedule A) classification. After studying the Bureau of the Census' cross-reference volume,[4] and after several conversations with the Bureau, it was determined that these seven categories were each virtually identical as between the Schedule A and the Schedule B classification schemes. The seven commodities, along with their Schedule A and Schedule B classification numbers, appear in Table A5.2.9. Note that all seven commodities are included in Section 7 of the 1967 SITC product classification.

In order to insure continuity in the content of each commodity category, we referred to the bulletins published annually in the Census Bureau's _Changes in Schedule B_ and _Changes in Schedule A_. The most significant of these variations in the classification codes are indicated in Table A5.2.9.

New vs. Control Commodities. The criteria used in placing commodities into the new category were, of necessity, somewhat loosely defined. This is due to a lack of comparable classification of United States imports and exports. We had hoped to have large samples of both new and control commodities, with the determination of category to be based on first production dates. Lacking this large sample, we used all available evidence, both direct and indirect in attempting to classify our sample of seven commodities.

The considerations lying behind the assignment of each commodity into either the new-commodity or control-commodity group as of 1952, the date when our study begins, are presented below.

"New" Commodities:

Standard Electric Typewriters. Our choice with respect to this commodity was predicated largely by product age. IBM introduced the first electric typewriter in the late 1940's.

Table A5.2.9

Commodity Sample

Commodity	Schedule A			Schedule B	
	a	b	c	d	e
Std. Electric Typewriters	777100	77710	714:1010	*	7785-100
T.V. Tubes & Parts	707812 707815	70828 70832	729:3015 729:3010	7096-200	7100-030
Std. Non-Electric Typewriters	777000 777200	77700 77720	714:1020 714:1030	7785-000	7785-300 7785-400
Refrigerators (electric)	705715	70580	725:0110	7100-990	7070-100
Electric Household Vacuum Cleaners	706910	70691	725:0320	7069-000	7069-010
Electric Ranges	707200	70720	725:0505	7090-850	7090-850
Radio-Phonograph Combinations	707705	7083	724:2020	7100-900	7100-230

[a]1957 six-digit classification.

[b]1958 five-digit classification.

[c]1956 seven-digit classification.

[d]Pre-1960 seven-digit classification.

[e]1960 seven-digit classification.

*Not established prior to 1960.

Table A5.2.10

Date of Establishment of Each Classification

Commodity	Schedule B	Schedule A
Std. Electric Typewriters	1952	1960
T.V. Tubes & Parts	1952	1956
Std. Non-electric Typewriters	1926	*
Refrigerators (electric)	1930	1954
Electric Household Vacuum Cleaners	a	a
Electric Ranges	1933	1944
Radio-Phonograph Combinations	1949	1954

*Not available.

[a]Classification existed prior to 1920.

The U.S. Department of Commerce first established an export classification for standard electric typewriters in 1952 (See Table 5.2.10). Therefore in 1952, the first year of our study, this commodity had barely appeared on the international commodity market. We elected therefore to include this commodity in the new category.

Standard Non-electric Typewriters. Since Remington first produced a typewriting machine in the late 1870's some support of assigning this commodity to the new product category, on a basis other than product age -- in years -- must be tendered. Our choice was predicated primarily on the introductory remarks appearing in a recent Business and Defense Service Administration study of world trade in standard, non-electric typewriters:[5]

> Because it is primarily a tool of commerce, the typewriter and the technology of its manufacture have remained until recently, of special interest only to highly developed industrial countries, . . . (but) significant changes have occurred in the pattern of world trade in typewriters in recent years. Several European countries that had never before produced typewriters, have become important producers and exporters. Also, limited typewriter assembly and production operations have been established in Latin America and in Asia.[5]

T.V. Tubes and Parts. R.C.A., launched television commercially in 1939, and this commodity first entered the U.S. trade classification (Schedule B) in 1952. While several European electronics firms have (as of 1965) done successful development work on tubes and sets, and have benefited from cross-license agreements with U.S. firms; master patents, as of 1965 were still held by R.C.A., and the flow of technical payments has been mainly in one direction.[6]

Television sets and television technology is therefore, still a "new" one in 1952, and is dominated by the U.S. We felt reasonably well assured then, that the correct placement of television was in the new-product category.

Refrigerators. A "new" commodity should not exhibit high ownership as a percent of total households in a relatively developed economy such as the U.K. In a study of long-termed demand for consumer durables by the National Institute, data is presented showing that ownership of refrigerators in 1952 as a percent of total households was only 3.9%; by 1966 ownership had climbed to 47.0% of total households, and to 82% of upper-class households.[7]

Secondly, while an export classification for refrigerators was first established in 1930, it was not until 1954, that there were sufficient imports of refrigerators to warrant establishing an import classification.

Refrigerators, then, were included among the "new" commodities. But this choice was more due to demand than supply considerations; in fact, it is our opinion that refrigerators do not strictly belong to either category, but due to aforementioned demand considerations, we expected its dynamic trade behavior to reflect that of a new commodity.

Control Commodities

The remaining three commodities in our sample were assigned to the control group, largely on the basis of age. Both export and import classifications for vacuum cleaners appear in the Department of Commerce classification in the early 20th century. We have been importing electric ranges since the end of World War II.

Radio-phonograph combinations do not appear in the export classification until 1949, but embodied in this commodity, is a technology that was widely diffused by 1952: sound transmission via vacuum tubes, an invention of the early 1920's. As evidence for this, one need only take notice of the vigorous market which exists today for early 20th century American and European vacuum tubes.

Notice must be taken here of the transistor revolution of the late fifties. Since it was initiated by the U.S., we would expect that the cross-sectional behavior of the jth U.S. bilateral trade balance in radio-phonograph combinations in the early-to-mid sixties to exhibit some of the properties of that of a new commodity.

3. Data Sources

1. First trade dates: U.S. Department of Commerce, Bureau of the Census, Schedule B, Statistical Classification of Domestic and Foreign Commodities Exported from the United States: January 1, 1958 Edition, Superintendent of Documents, Washington, D.C., reprinted January 1, 1962.

2. Relation between seven-digit 1965 SITC classification and 1964 six-digit classification: U.S. Department of Commerce, Bureau of the Census, Changes in Schedule B ________,: January 1, 1965 Edition, Public Bulletin B-1; Superintendent of Documents, Washington, D.C., May, 1965.

3. Export Data: U.S. Bureau of the Census, United States Exports, Commodity by Country; (FT 410), 1952 Edition through 1968 Edition, Washington, D.C.

4. Import Classification: U.S. Bureau of the Census, Schedule A, Statistical Classification of Commodities Imported into the U.S., 1952 through 1968 Editions, Washington, D.C.

   AND: ________, Changes in Schedule A, . . . , 1952 through 1958 Bulletins, Washington, D.C.

5. Cross Reference: U.S. Department of Commerce, Bureau of the Census, U.S. Foreign Trade Statistics, Classification and Cross-classification, 1968 Edition, Superintendent of Documents, Washington, D.C.

6. Import Data: __________, United States Imports for Consumption, Commodity by Country of Origin, 1952 Edition through 1968 Edition, (Ft110, Ft120, Ft125), Superintendent of Documents, Washington, D.C.

7. Income Data: Agency for International Development, Statistics and Reports Division, Office of Program and Policy Coordination, Gross National Product, Growth Rates and Trend Data, By Region and Country, Washington D.C., April 25, 1969.

FOOTNOTES

[1]This example is from, Carl F. Christ, Econometric Models and Methods, pp. 388-389.

[2]Ibid., p. 389.

[3]Sources: 3,6.

[4]Sources: 5.

[5]World Trade in Typewriters 1948-1958, B.D.S.A., Washington, D.C., June, 1959, p. 1.

[6]"Research and Development in Electronic Capital Goods," National Institute Economic Review, #34, November, 1965.

[7]"Long-term Forecasts of Demand for Cars, Selected Consumer Durables, and Energy," National Institute Economic Review, #40, May, 1967.

BIBLIOGRAPHY

Articles

Arrow, Kenneth J., "Induced Technical Change and Patterns of International Trade, Comment," The Technology Factor in International Trade, pp. 128-132.

Balassa, Bela, "Country Size and Trade Patterns, "American Economic Review, March, 1959, pp. 201-204.

Bertrand, T.J. and Vanek, J., "Growth with Technological Change, Variable Returns to Scale and a General Saving Function," La Rivista Internazionale de Science Economiche e Commerciali, 1969, pp. 751-753.

Chipman, "Introduced Technical Change and Trade Patterns," The Technology Factor in International Trade, pp. 95-127.

Freeman, C., "Long-Term Forecasts of Demand for Cars, Selected Consumer Durables, and Energy," National Institute Economic Review, May, 1967.

__________, "Plastics Industry: A Comparative Study of Research and Innovation," National Institute Economic Review, Nov., 1963.

__________, "Research and Development in Electronic Capital Goods," National Institute Economic Review, November, 1965.

Gruber, W., Mehata, and Vernon, R., "The R&D Factor and International Investment of U.S. Industries," Journal of Political Economy, February, 1967, pp. 20-37.

Gruber, W. and Vernon, R., "The Technology Factor in a World Trade Matrix," The Technology Factor in International Trade, pp. 233-272.

Hirschman, A.O., "An Invitation to Theorizing on the Dollar Glut," Review of Economics and Statistics, February, 1960, pp. 100-102.

Hufbauer, G.C., "The Impact of National Characteristics and Technology on the Commodity Composition of Trade in Manufactured Goods," The Technology Factor in International Trade, pp. 145-232.

Houthakker, H.S., and Magee, S.P., "Income and Price Elasticities in World Trade," Review of Economics and Statistics, May, 1969, pp. 111-125.

Jones, Ronald W., "The Role of Technology in the Theory of International Trade," Mimeo.

Keesing, D.G., "The Impact of Research and Development on U.S. Trade," Journal of Political Economics, February, 1967, pp. 38-48.

_______, "Labor Skills and Comparative Advantage," American Economic Review, Papers and Proceedings, May, 1966, pp. 249-258.

Kravis, Irving B., "'Availability' and Other Influences on the Commodity Composition of Trade," Journal of Political Economics, April, 1956, pp. 134-155.

Lancaster, Kelvin, "Change and Innovation in the Technology of Consumption," American Economic Review, Papers and Proceedings, May, 1966, pp. 14-23.

_________, "A New Approach to Consumer Theory," Journal of Political Economics, April, 1960, pp. 132-157.

Posner, M.W., "International Trade and Technical Change," Oxford Economic Papers, October, 1961, pp. 323-341.

Quandt, R.E., "The Construction of Travel Demand Models with Incomplete Data," Mimeo.

Stobaugh, Jr. R.B., "Where in the World Should We Put That Plant?", Harvard Business Review, January-February, 1969, pp. 129-136.

Vernon, R., "International Investment and International Trade in the Product Cycle," Quarterly Journal of Economics, May, 1966, pp. 190-207.

Wells, Jr., Louis R., "Test of a Product Cycle Model of International Trade: U.S. Exports of Consumer Durables," Quarterly Journal of Economics, February, 1969, pp. 152-162.

Books

Balassa, Bela, Trade Liberalization Among Industrial Countries, Objectives and Alternatives, New York: McGraw-Hill, 1967.

Christ, C.F., Econometric Models and Methods, New York: John Wiley & Sons, 1967.

Fei, J.C., and Ranis, G., Development of the Labor Surplus Economy, Homewood, Illinois: Richard D. Irwin, Inc., 1964.

Hirsch, Seev, Location of Industry and International Competitiveness, Oxford: Clarendon Pres, 1967.

Hoffmeyer, Erik, Dollar Shortage, Amsterdam: North-Holland Publishing Company, 1958.

Hufbauer, G.C., Synthetic Materials and the Theory of International Trade, Cambridge: Harvard University Press, 1966.

Johnson, H.G., Comparative Cost and Commercial Policy Theory for a Developing World Economy, Wicksell Lectures, 1968; Sweden: Almqvist & Wicksell, 1968.

Kindleberger, C.P., The Dollar Shortage, New York: McGraw-Hill, 1950.

____________, American Business Abroad, New Haven: Yale University Press, 1969.

Klein, L., An Introduction to Econometrics, Englewood Cliffs, New Jersey, Prentice-Hall, 1962.

Maizels, A., Industrial Growth and World Trade, New York: Cambridge University Press, 1963.

McGraw-Hill, Department of Economics, Foreign Operations of U.S. Industrial Companies, 1960-1961, New York: McGraw-Hill, 1961.

O.E.C.D., The Gaps in Technology Between Member Countries, General Report, Paris: 1969.

_______, Gaps in Technology: Computers, Paris: 1969.

O.E.C.D., Gaps in Technology: Scientific Instruments, Paris: 1968.

_______, Gaps in Technology: Plastics, Paris: 1969.

Vernon, R., ed., The Technology Factor in International Trade, New York: Columbia University Press, 1970.

Documents

Agency for International Development, Statistics and Reports Division, Office of Program and Policy Coordination, Gross National Product, Growth Rates and Trend Data, By Region and Country, Washington, D.C. April, 1969.

Business Defense and Services Administration, World Trade in Typewriters 1948-1958, Washington, D.C., June, 1959.

_______, Changes in Schedule A, Statistical Classification Commodities Imported Into the U.S., Bulletins 1952 through 1958, Washington, D.C.

_______, Schedule A, Statistical Classification of Commodities Imported into the U.S., 1952 edition through 1958 edition, Washington, D.C.

_______, Schedule B, Statistical Classification of Domestic and Foreign Commodities Exported from the United States: January 1, 1965 edition, Washington, D.C., May, 1965.

_______, U.S. Foreign Trade Statistics, Classification and Cross-Classification, 1968 edition, Washington, D.C.

_______, United States Exports, Commodity by Country; (FT 410), 1952 edition through 1968 edition, Washington, D.C.

NOTES